"This collection of essays passionately argues for a creative hope to confront the catastrophism of today. In the face of pending global destructivity, each author thoughtfully explores how psychoanalysis can authentically challenge the drastic warnings of our time."

Jan Abram, *President, European Psychoanalytical Society; author,* The Surviving Object: Psychoanalytic clinical essays on psychic survival-of-the-object

"To read these essays is to engage in an exciting conversation with eminent psychoanalysts and academic scholars on the global impact of rapidly increasing climate change affecting every aspect of our existence. This book invites readers to join in a new narrative that acknowledges the truth but also offers hope for next generation stakeholders."

Bonnie E. Litowitz, *Ph.D.; Editor-in Chief Emerita,* Journal of the American Psychoanalytic Association (JAPA); *editor (with Glen Gabbard and Paul Williams),* Textbook of Psychoanalysis, *2nd faculty, Chicago Psychoanalytic Institute, retired*

"If we needed an original, direct, nuanced psychoanalytic inquiry into the dangers of catastrophic thinking, this book is it. Cosimo Schinaia has brought together a first-rate group of authors to engage in disentangling catastrophism from creativity, radical hope, essential inter-dependence and psychic growth through a series of dazzling essays. This book is of great relevance to us all."

Paul Williams, *member, Psychoanalytic Institute of Northern California; Joint Editor-in-Chief,* The International Journal of Psychoanalysis, 2001–2007

Against Catastrophism

Against Catastrophism explores catastrophism from multiple vantage points and considers the impact of ongoing crisis on individuals.

Bringing together contributors from psychoanalysis, economics, anthropology, and gastroenterology, this book explores themes including fossil fuel culture, social movements like Extinction Rebellion, the COVID-19 pandemic, media messaging, and the future of food supply chains. By assessing the value of a constant barrage of information about catastrophes and considering the need for a containing environment, the chapters explore how we can avoid endorsing a closed-off vision of the future and instead unlock possibilities. The book concludes with a discussion of optimism, radical hope, and how we can put forward a new narrative on nature.

Against Catastrophism will be of great interest to psychoanalysts, psychologists, psychiatrists, economists, anthropologists, sociologists, food scientists, environmentalists, ecologists, politicians, and communication experts.

Cosimo Schinaia is a training and supervising psychoanalyst of SPI (Italian Psychoanalytic Society), a full member of the IPA, and former director of the Department of Mental Health in Central Genoa. He received the IPA Climate Award in 2023 for his interests and studies on the relationships between psychoanalysis and ecology. His books *On Paedophilia; Psychoanalysis and Architecture; Psychoanalysis and Ecology*; and *Outsider Art and Psychoanalytic Psychiatry* are also published by Routledge and have been translated into many languages.

Against Catastrophism

Climate Change, Pandemics, and Hope for the Future

Edited by
Cosimo Schinaia

Routledge
Taylor & Francis Group

LONDON AND NEW YORK

Designed cover image: Getty images / piyaset

First published 2025
by Routledge
4 Park Square, Milton Park, Abingdon, Oxon OX14 4RN

and by Routledge
605 Third Avenue, New York, NY 10158

Routledge is an imprint of the Taylor & Francis Group, an informa business

British Library Cataloguing in Publication Data
A catalogue record for this book is available from the British Library

ISBN: 978-1-032-81200-7 (hbk)
ISBN: 978-1-032-81198-7 (pbk)
ISBN: 978-1-003-49860-5 (ebk)

DOI: 10.4324/9781003498605

Typeset in Times New Roman
by Taylor & Francis Books

Contents

Figures

Contributors

Cosimo Schinaia is a Psychiatrist, former Director of Mental Health Department of Central Genoa, a Training and Supervising Psychoanalyst of the Italian Psychoanalytic Society (SPI), and a Full Member of International Psychoanalytical Association (IPA). He lives in Genoa, Italy, where he works in private practice. His latest books are *Psychoanalysis and Ecology* (Routledge, 2022) and *Outsider Art and Psychoanalytic Psychiatry* (Routledge, 2024), translated into several languages. He is co-author with Karyn Todes of the chapter "Psychoanalysis and climate change", in G. Gabbard, B. E. Litowitz and P. Williams (Eds), *Textbook of Psychoanalysis* (Washington, DC: APA, 2024).

Orazio P. Attanasio is the Cowles Professor of Economics at Yale, a Member of the American Academy of Arts and Sciences, and a Fellow of the British Academy and of the Econometric Society, of which he was President in 2020. In 2014 he served as President of the European Economic Association. He lives and works in New Haven, USA.

Luca Caldironi is an MD and clinical Psychiatrist, member of the Italian Society of Psychoanalysis (SPI), American Psychoanalytic Association (APsaA) and of the International Psychoanalytical Association (IPA), on the Board of the Confederation of Independent Psychoanalytic Societies (CIPS) and of The International Association for Art and Psychology (IAAPs). He lives and works in Venice, Italy.

Christine Franckx is a Child Psychiatrist and Training Psychoanalyst for adults, children and adolescents with the Belgian Psychoanalytic Society (BPS), a full member of the IPA, a former President of the BPS (2016–20), the founder of Psychoanalytic Psychotherapy Center Infant 3 PSY (ages 0–6) and of the non-governmental organization GiO, which promotes psychoanalytic thinking and practice in the community. She is the co-editor of *Eros op de scene* (2021, with Joyce McDougall) and co-editor of *Kinderlijk Trauma* (2023, with Sandor Ferenczi). She lives and works in Antwerp, Belgium.

Attilio Giacosa is Professor of Gastroenterology at the Centro Diagnostico Italiano in Milan, He has previously been Director of the Gastro-enterology and Nutrition Unit at the National Cancer Institute of Genoa and Adjunct Professor of Gastroenterology at the University of Genoa. He lives in Pavia and works in Genoa, Italy.

Mark Halle is an Environmental Policy specialist and the Founder of Better Nature, an organization focused on a narrative shift in the fields of development and environment. He was the former Executive Director of the International Institute for Sustainable Development – Europe from 2002 until 2016. He is also Senior Advisor to NatureFinance and to the international network of Financial Centres for Sustainability (FC4S). He lives and works in Geneva, Switzerland.

Gohar Homayounpour is a Full Member of the American Psychoanalytic Association and the IPA. Her first book, *Doing Psychoanalysis in Tehran* (MIT, 2012) won the Gradiva Award. Her latest book is titled *Persian Blues: Psychoanalysis and Mourning* (Routledge, 2022). She is the founder of the Freudian group of Tehran and a member of the scientific board of the Freud Museum in Vienna. She lives and works in Tehran, Iran and Paris, France.

Ronny Jaffè is Training and Supervising Analyst of the SPI and a full member of the IPA. He is the President of the Italian Psychoanalytic Society. He lives and works in Milan, Italy. He is the co-author with B. Reith et al. of *Initiating Psychoanalysis* (Routledge, 2012) and *Beginning Analysis* (Routledge, 2018) and with D. Campbell of *When the Body Speaks* (Routledge, 2022).

Alfredo Lombardozzi is Training and Supervising Psychoanalyst with the SPI and a full member of the IPA. He was previously Professor in Psycho-analytic Anthropology at the University in Chieti and L'Aquila. He is Editor in Chief of the *Rivista di Psicoanalisi*. He lives and works in Rome, Italy.

Mauro Van Aken is an Anthropologist and Associate Professor at the Uni-versity Milan-Bicocca, where he teaches Environmental Anthropology and Climate Crisis. He lives in the region of rural Pavia and works in Milan, Italy. He is author of *Campati per Aria* (Down to Air), (Milan: Eleuthera, 2020) and the chapter "Palestinian memories and practices of weather relatedness", in P. Silitoe (Ed), *The Anthroposcene of Weather and Climate* (New York: Berghahn Books, 2021).

Preface

Cosimo Schinaia

Catastrophism is the French naturalist Georges Cuvier's theory asserting that changes on Earth occur through sudden, violent, and irreversible upheavals. This doctrine posits that Earth's history has been shaped primarily by short-lived cataclysmic events rather than gradual processes acting over extended periods of time. Coined in 19th-century geology, the term "catastrophism" contrasts with the principle of "actualism," according to which the present is the key to the past, thus the same processes and natural laws prevailed in the past as those we can now observe or infer from observations and with the concept of "uniformitarianism," which suggests that geological changes happen gradually and uniformly.

Over time, the term has come to define the tendency to magnify the probability and severity of a potential threat. The American Psychological Association dictionary defines catastrophism as exaggerating the negative consequences of events or decisions. People are said to be catastrophizing when they think that the worst possible outcome will occur from a particular action, or in a particular situation, or when they feel as if they are in the midst of a catastrophe in situations that may be serious and upsetting but are not necessarily disastrous.

Susan Sontag wrote in 1965 that we live in an age of extremity, under continual threat of two equally fearful, but seemingly opposed, destinies: unremitting banality and inconceivable terror. For one job that fantasy can do is to lift us out of the unbearably humdrum and to distract us from terrors, real or anticipated – by an escape into exotic dangerous situations which have last-minute happy endings. But another one of the things that fantasy can do is to normalize what is psychologically unbearable, thereby inuring us to it. In the one case, fantasy beautifies the world. In the other, it neutralizes it. Her words could very well also have been written in the current era.

According to Fredric Jameson (1991,) an important theme of postmodern representation is an inverted millenarianism, in which premonitions of the future, catastrophic or redemptive, have been replaced by senses of "the end of this or that," so the future is already our past and the end we are waiting for has already happened.

Paul Hoggett (2023) writes that the danger is that we switch from denial to terror, from complacency to catastrophe and that "end of the world" feeling.

Obviously, it is neither helpful nor appropriate to use the word "catastrophism" as a synonym for the scientifically documented denunciation of the environmental dangers we face. Ecological ethics and science posit that there are good reasons to be alarmed. The speed of the changes that human activity has brought upon the earth's climate, oceans, soils, and fauna is progressively depriving us of the planet we once knew.

We cannot be ingenuously optimistic or irresponsibly indifferent toward the limitations of the Earth's natural resources, although we too often consider them unlimited. Unconsciously identifying ourselves with what we perceive as omnipotent and immortal technology, we do not want to see the dark side of our societal wellbeing and Western lifestyle; we protect ourselves against intolerable feelings of insignificance, deprivation, loss, fear of death, and the sense of guilt that would result from acknowledging our implicit connivance or cohabitation with the blind exploitation of natural resources, the underestimation of the costs, and destructive repercussions that arise from it, reacting with a severe and pervasive apathy. Denialism involves the organized and planned broadcasting of scientific fake news for commercial, political, or ideological aims. It leans on the individual defense mechanisms, such as negation and disavowal (Weintrobe 2013).

Using the same language as right-wing climate science deniers, who fetishize industrial society by rendering invisible the ecological conditions on which it depends and minimize the extent of the collateral damage caused by the environmental impact or contest the value of scientific discovery, should give the impression that the dangers are non-existent or exaggerated. Putting accurate and motivated environmental warnings in the same category as apocalyptic fundamentalism actually leads to underestimating the threats we face and diverting efforts from the need for an effective counter-offensive against the existential climate crisis.

The way out of both catastrophism and the sense of transience and the bubble of denialism leads to mourning the world that is ending, with its promises and certainties, and preparing to inhabit the complexity of the new world, to reopen the horizon of possibilities and the future, without abandoning but re-signifying and complexifying the founding ideas of modernity: progress, science, humanism, and universalism.

At times, we psychoanalysts have tended to refer to a mythical and mythicized inner world, of which we were competent and exegetes, belonging to a human being disconnected from the landscapes to which he or she gave shape, from the vicissitudes and contingencies of real history, from the fluctuation of its frontiers. After a phase of idealization of the separateness of the internal world, which was a reaction to the positivistic culture that gave no room for emotional-affective determinations in the observation of facts, we have, however, increasingly come to realize that knowledge and care of the internal require both the abandonment of preclusive scotomizing skills and a curious attention to – and thoughtful care for – the external, without cuts, without caesuras.

The external territory is internal for Freud, who in his "New Introductory Lectures on Psychoanalysis" (1933) designates the repressed as internal foreign territory to the ego, just as reality is external foreign territory, but the internal territories also invade and transform the composition, partition and boundaries of the external territories in a continuous and endless work of osmosis. We live in a time when the large and overt crises of external reality we analysts share with our patients get mixed up with the subjective and intimate crises of their internal world.

The consequence is the need for fruitful and propulsive contaminations among the different disciplines that, in the full respect of their scientific and epistemological statutes. aspire to constitute a new way of thinking, one that would be unstable and incomplete, as well germinal and transformative. Despite its enormous and well-defined body of work, a psychoanalysis is increasingly taking shape that is still able to enrich itself thanks to theoretical and clinical terms derived from other cultures. Thus, psychoanalysis can become a living, evolving organism that is able to understand and imagine which humanity is taking form or, rather, which humanity we are constructing (Preta 2019).

This book stems from my request to five psychoanalysts, two economists, an anthropologist, and a gastroenterologist who is an expert in food science to try their hand at different aspects of catastrophism from multiple vantage points.

Chapter 1: "Hope in a Changing World between Individual Creativity and Collective Working Through" by Cosimo Schinaia.

In "On Transience" (1916), Freud describes the anticipatory mourning of the object, which consists of the rejection of its cathexis to avoid suffering from its loss and suggests that the environment and affectively invested objects can be experienced in a peculiar atmosphere of incipient loss and impending fear of the end. Referring to Freud, the need to react to climate terrorism by opposing a closed-off vision of the future and making room for possible futures is highlighted. The need for a mature and depressive working hope is emphasized that, through the integration of individual creativity with collective elaboration, skirts the childish notion of "all and now" and promotes firm and enduring anti-entropic commitment against the forces pushing towards the dissolution of bonds.

Psychoanalysts can play a crucial role in cultivating the ability to think and dream a better future and contribute to valuing a sense of measure and sobriety, proposing a good enough life, in which there can be room for love and creativity, counteracting magical and illusory thinking, and looking with sincerity and integrity even at the negative aspects of existence.

Chapter 2: "Radical Hope" by Gohar Homayounpour.

This chapter attempts to elaborate the problematics of catastrophism in the face of excruciating social and political realities in Iran. Following the onset of the radical feminist uprising in Iran that has been ongoing since September 16, 2022, the author's wish is to illustrate that the antidote to the catastrophism of our times can be witnessed in the resurrection of the erotic, a resurrection that has been at the very core of this subversive feminist revolt, of a birth of new feminine epic hero, towards an ethics of life, and its conditions.

In the very fabric of the saying "woman, life, freedom," there is a clear connection to life, binding, linking, libido and sublimation. They are not saying we want to die for freedom, they are saying we want to live for freedom, they are taking to the streets, risking their lives for better conditions of life. A life of dignity, pleasure, freedom, and of passionate transformations – all the very derivatives of the life drive, the ethics of the erotic, indeed the enlivening antidote to catastrophism in the most devastating of conditions.

Chapter 3: "A Changing World: Psychoanalysis between Catastrophe and Hope" by Alfredo Lombardozzi.

The author argues that we are in the midst of processes that favor forms of forgetting and denial, both on an individual and societal level, as a defensive reaction to problems that are unaddressable, or perceived as such.

This leads to favoring a heightened ideological polarization of opinions rather than encounter and dialogue, or to rigid positions that re-propose conflicts, as is the case with climate change between those who deny the existence of the problem and the responsibilities of human intervention, and those who instead propose radical solutions that do not take into account the limits within which one can still act by setting reasonable goals. In the opening of a dialogue in the transition between the different figures of catastrophe, a space is created – the ground in which the figure of Hope takes shape.

Chapter 4: "Catastrophic Fossil Culture and Other Desires of Relatedness" by Mauro Van Aken.

Climate crisis stands as a symbolic failure of the cultural imagination: the notion of nature as separated from societies has been fossilized by the carbon economy and represents a cosmology, a set of representation of time, of the future, of human beings and of nonhumans as at disposal and separated. This engenders the uncanny aspects of the Western notion of "nature" in the Anthropocene: living agents show their interrelations with the human condition, act, and re-emerge within "emergencies" in "catastrophic" scenarios.

The carbon economy shifts today from a promise of modernity and secular salvation to its destructive character in defining futures: it is enmeshed in a petro-culture, a symbolic set of meanings in full crisis. A decarbonization of the economy relies on a decarbonization of the imagination. A new social and cultural alphabet in defining nonhuman relations is underway, often with a strong generational character, based on metaphors of relatedness and interdependencies, and on the elaboration of a new "crisis of presence," as ritualized patterns in giving meaning to radical changes, founded on desires of relations and even, of limits.

Chapter 5: "Illusory Immunity and Actual Inhumanity" by Ronny Jaffè.

Referring to the social writings of Freud, to Bion, and other authors on group dynamics, the author focuses on the concept of illusory-delusional immunity in respect to environmental changes. Illusory-delusional immunity can match the inhumane cruelty towards others and towards nature, creating states of panic, violence, indifference.

The sense of limitation that characterizes human beings can be obscured by mass fantasies of an illusory-delusional nature of technological and scientific omniscience and, as a result, turn into an omnipotent and borderless ideology with all the attendant risks. Being able to introject a sense of limitation and be able to do a reality check on the habitat in which we live could allow for behaviors compatible with sustainability by getting out of both false illusions and a gloomy resignation, where one accepts that Cassandra's warning is an inexorable truth by which, in the meantime, everything is allowed, without vital planning for future generations.

Chapter 6: "Catastrophe versus Catastrophic Change. Between Psychoanalysis and Art" by Luca Caldironi.

In a world where the term "catastrophe" is increasingly used as a signal of an impending "apocalypse," the author believes it is important to recover its etymological meaning.

This digression allows the author to approach the "threat" of a universal catastrophe by coupling it with that proposed by Bion of "catastrophic change". With this concept, Bion glimpses the possibility of also taking on a positive meaning. He sees the possibility of a phenomenon that marks an evolutionary leap, both regarding individual mental growth, as well as group dynamics and social transformations. The author stresses the value of creativity and the artistic moment in the compass of our navigating by sight.

These considerations seem profoundly current and useful in a civilization, as in the one in which we are living, that tends to deny death and its mystery. Furthermore, they seem useful and prevent us from disavowing, therefore denying, the problem of the possible progressive destruction of our planet, and its vulnerability, which, like ours, can be denied.

Chapter 7: "Climate Change and Adolescence: A Dangerous Collusion of Internal and External Catastrophe" by Christine Franckx.

The ecological crisis impacts adolescent lives worldwide and creates a specific risk for youngsters of becoming trapped in their development by the superposition of two coinciding catastrophic processes. On the one hand, the internal libidinal crisis, launched by puberty, has the quality of a mental catastrophe and is always potentially traumatic if the psyche does not succeed in reconciling the forces of life with the death instinct. On the other hand, the geopolitical catastrophe of global warming makes it difficult for adolescents to find the appropriate parental and societal guidance to contain the anxieties this collective threat provokes to transform the illusion of infantile omnipotence in the face of the real danger of extinction.

Not only are adolescents pessimistic about the uncertain future of the environment, they may also feel anxiously persecuted by their primary objects who failed to protect them sufficiently. This new global geo-socio-political emergency situation therefore impacts the development of their ego-functions with a risk for a fragile constitution of their adult self. In the current situation of climate change there simply is no representation at all.

Chapter 8: "Birth is not Destiny" by Orazio Attanasio.

One possible interpretation of catastrophism is that humans are not in full control of their destiny, and that, at least after some point in life, subsequent development cannot be altered in a substantial fashion. Of course, that might be particularly desperate for those who have drawn the short straw in the lottery of life but elating for the lucky ones. However, from a societal point of view, such a description of society is dystopic.

It is now well recognized that what happens in the early years of life (the first thousand days since conception), has long-run consequences that are visible in a variety of outcomes. And many of the events that imply good or bad lack cannot be controlled. At conception, different individuals draw different combinations of genetic material. And immediately after, the phenotypes generated by these combinations are exposed to a variety of environmental factors that shape their expression. Recognizing that early childhood development can be affected by the environment in which children grow up shows that initial disadvantages can be compensated. More importantly, the fact that early disadvantage, while not being controlled by the individuals that experience it (the children), can be changed draws attention to the moral necessity of designing policies that can allow the full development of every individual.

Chapter 9: "The Food of Tomorrow" by Attilio Giacosa.

Food has always been an essential element in human life, both for its crucial role in ensuring physical survival and for its emotional, social and cultural implications.

The ongoing demographic evolution, the profound climate changes we are witnessing, and the consequences of recent critical human behavior throughout the food chain, from agricultural production to industrial processing, packaging, transport, and final consumption, impose drastic changes of course in order to avoid societal and environmental collapse. The "zero hunger" objective is a categorical obligation in a world that today where the number of obese people is just as great as the number of those suffering from hunger on a global scale.

We need to produce better and eat better to promote the health of humans and the planet. We need to revisit the agricultural world by making it more sustainable. The type of food of tomorrow needs to be rethought. One approach will be to favor plant-based foods. Farming livestock will have to be restricted over time: less land will have to be devoted to raising animals, and new sources of protein will be needed to replace the current ones, which are more ecological and less wasteful of environmental resources. The future will see the growing cultivation of meat, fish, milk, and dairy products in the laboratory, derived from animal stem cells. The use of insects as an optimal protein source will become an obligatory choice. Therefore, our relationship with food needs to be rethought. Traditional, innovative, healthy, sustainable, responsible, and at the same time pleasant must be the adjectives that inspire our food choices, to promote the quality of our lives and the health of the planet.

Chapter 10: "Pushing Back on Catastrophism: The Case for a New Nature Narrative" by Mark Halle.

A narrative is a frame within and around which stories are told. The narrative shapes how the stories are heard and how issues are approached and apprehended. A narrative that comes to dominate carries with it an aura of self-evident truth and a sense of inevitability, as if everything is beginning to converge towards the goal that is the narrative's central message. The narrative – and the stories that compose and articulate it – begin to build a momentum of change until tipping points are reached, and a new reality is born.

The dominant narrative in respect of nature has been one of impending catastrophe. The thinking behind this narrative – that people can be frightened into acting – is flawed.

Nature and ecosystems – uniquely – possess the capacity to regenerate, to restore biodiversity and diverse, complex ecosystems. But this will require a fundamental shift from the current, catastrophist narrative to one that is mobilizing, hopeful, and exciting. It will need to play to the non-material values we all cherish and underline how nature lies at the foundation of everything we as human beings want.

Chapter 11: "Catastrophism and Media Catastrophic Images" by Cosimo Schinaia.

Ferenczi's concept of "the identification with the aggressor" developed in working with traumatized patients, and Ogden's concept of the "autistic-contiguous position", can allow us to get in touch with the most common attitudes and beliefs in exploring individual and group defense mechanisms against climate change and environmental disasters. Using psychoanalytical knowledge, we can try to help people to face the reluctance to fully acknowledge the severity of climate change and thus to change harmful behaviors in our relationship with the non-human world.

The use of abusively repeated terrifying and bombarding images on television and the Internet is criticized. These serial images purport to inform us about devastation and disaster around the world, but in effect they distort and paralyze psychic functioning. There is a risk that climate terrorism could foster the emergence of persecutory and primitive anxieties and even the activation of psychotic defenses.

The other risk that people incur is that of not authentically and depressively being in touch with their own deep-seated anxieties and of removing from themselves the sense of responsibility and awareness of their own participation in the creation of the damage being wrought on the environment.

References

Freud, S. (1933). The Introductory Lectures on Psychoanalysis. *SE* 22: 1–182.

Hoggett, P. (2023). *Paradise Lost? The Climate Crisis and the Human Condition.* Simplicity Institute Publishing.

Jameson, F. (1991). *Postmodernism or, the Cultural Logic of Late capitalism.* London: Verso.

Preta, L. (2019). *The Brutality of Things: Psychic Transformations of Reality.* Milan: Mimesis International.

Sontag, S. (1965). "The imagination of disaster". In: *Against Interpretation and Other Essays* (pp. 209–225). New York: Farrar, Straus and Giroux.

Weintrobe, S. (2013). "Introduction". In: S. Weintrobe (Ed), *Engaging with the Climate Change: Psychoanalysis and Interdisciplinary Perspectives* (pp. 1–15). London & New York: Routledge.

Chapter 1

Hope in a Changing World between Individual Creativity and Collective Working Through

Cosimo Schinaia

Introduction

The collapse of human civilizations is not a new subject. Anthropologist Ernesto De Martino (1977) remarks on the ubiquity of apocalypticism throughout history, noting that the theme of the world's end is as ancient as the world itself. "Apocalypse," "the end of the world," "millennialism," "millenarianism," and "fin de siècle" are all terminologies of ending: of life, epochs, the world, and the universe.

Catastrophist views have illustrious examples in earlier ancient religious and philosophical traditions, from China to India, Mesopotamia, Greek, and Roman civilizations (Star 2021) to the Middle Ages of the West: in particular, providential catastrophe, understood as divine intervention that produces catastrophe for the purpose of catharsis, the profound and radical transformation of an environment characterized by chronic problems, structural errors, and irresolvable evils. This is the case with the Flood, the Tower of Babel, the destruction of Sodom and Gomorrah, and the Apocalypse at the end of the New Testament.

However, the cultural significance, emotional tone, and contextual dynamics surrounding this theme vary across different eras, historical settings, social groups, and individuals, ultimately reflecting the forms of cultural coherence in which it participates. According to De Martino, there are two modes of representing the end of the world: one entails pure destruction, evoking the collapse of a cultural order without replacement, while the other constitutes a ritual dynamic of redemption from crisis, portraying it as a preparatory moment for the emergence of a new cultural order.

In July 1816 Lord Byron composed "Darkness," a poem about the end of the world in which the light of the sun, and mankind are finally extinguished. Universal darkness envelops everything, so that nothing is distinguishable any more, nothing is recognizable any more. The poet describes the return to the primordial, in whose deadly stillness the human being has collapsed along with his value system. Here are some lines of his poem:

DOI: 10.4324/9781003498605-1

> All earth was but one thought—and that was death,
> Immediate and inglorious; and the pang
> Of famine fed upon all entrails—men
> Died, and their bones were tombless as their flesh
> The meagre by the meagre were devoured

The year 1816 went down in history as the "Year Without a Summer" because of abnormal climate change in the previous year caused by the eruption of Mount Tambora on the island of Sumbawa (in present-day Indonesia), then part of the Dutch East Indies. Because of the most tremendous and powerful eruption in recorded human history, countless tons of volcanic ash with sulfur circulated in the upper atmosphere for years after the event, blocking out sunlight and lowering averages surface temperatures globally. In parts of North America and Europe temperatures dropped by more than eighteen degrees Fahrenheit. Consequently, great storms, rain, river flooding, and incessant summer snowfall occurred. It was even predicted that the sun would finally go out on July 18. This "prophecy" caused violence, riots, mass suicides, and manifestations of collective hysteria all over the world, and particularly in Europe.

David Levine and Matthew Bowker (2019) explore cultural and political trends centered around the belief that the world is a dangerous place unfit for human habitation, dominated by hate and destruction, and that our primary task is to survive by carrying on a life-long struggle against hostile forces. This concern lies not with the reality of existential threats but with the conviction that these threats exist, a belief that, while sometimes validated by real events, transcends reality-based sources of existential anxiety, fueling and shaping our experience of them.

When confronted with unfamiliar phenomena or paradigm shifts, alarm and anxiety often prevail, leading to defense mechanisms such as an inability to comprehend transformations or, worse yet, demonizing them without understanding (Preta 2023).

There exists a gap between the need for urgent, collaborative action related to the crisis, and the collapse of our mental functioning, given the distressing reality that we are the very agents of destruction. On the one hand, we are emotionally alerted and scared by the growing loss of biodiversity and the intensity of climate change, but on the other, we seem incapable of heeding these warnings amidst our daily routines, and fail to connect events and data, experienced as global and distant, with the urgency of change at the local level. In an attempt to quickly cover this gap, there is a serious risk of oversimplified theorizing. We risk uncritically embracing already known theories, also in psychoanalysis, in a highly complex situation that requires new thinking tools (Press 2019).

The Collapsology

In contemporary society, pessimistic and shocking literature permeates the postmodern narrative of decadence, manifesting in new forms and thriving on environmental risks increasingly viewed as certainties leading to the collapse of our civilization. The prevailing mantra is TINA (There is no alternative), a kind of ruthless self-fulfilling prophecy demanding deep adaptability without criticism and hope. Catastrophes are prophesied smugly, and new disciplines are founded with reference to the myth of the end.

Serge Latouche (2015a, 2015b) stresses the need of a "Pedagogy of disaster," according to which, when catastrophes are not too serious to destroy everything, but are serious enough to make people aware of the risk they run, they have a pedagogical role and help to decolonize the imaginary.

Today, however, catastrophic theories have gone further.

Collapsology is a neologism used by Pablo Servigne and Raphaël Stevens (2015) to designate transdisciplinary studies on the general collapse of societies induced by climate change, the resource scarcity, mass extinctions, and natural disasters. The title of the book "Another End of the World is Possible: Living the Collapse (and Not Merely Surviving It)" by Servigne, Stevens and Chapelle (2018) could be seen as the manifesto of collapsologists. The first author of reference for the collapsologists is Hans Jonas (1979) with the "prophecy of doom", to which one must pay more heed than to the prophecy of bliss, lest what is feared occur. The second one is Jean-Pierre Dupuy (2002), who points out the fragility of the precautionary principle, which maintains that we do not act in the face of catastrophe because we are not sure of knowing enough to act effectively. With his "enlightened doomsaying," Dupuy highlights how the problem is not scientific uncertainty but the impossibility of believing that the worst is about to happen. To prevent catastrophe, one must believe in its possibility before it happens.

For these authors, the breakdown of industrial civilization is a highly likely scenario, necessitating alignment of politics, society, culture, and spirituality with this new reality. Utopia may emerge from the coattails of ruin; in the aftermath of collapse, humanity at last could build a sustainable society. In their end-of-the-world narrative, the collapsologists explore the positive consequences of collapse, envisioning conversion into a richer spiritual, ethical, artistic, and emotional life where people would feel a greater connection with each other and the natural world. Such is the pessimism regarding reform in the pre-collapse world that collapse is seen as the most viable pathway to a sustainable planet.

In this regard, it is worth recalling how Naomi Klein in her *The Shock Doctrine* (2007) anticipated the critique of collapsology, describing the way global capitalism cynically exploits catastrophes (wars, political crises, natural disasters) to get rid of the "old" social constraints, and thus impose its agenda on the tabula rasa created by catastrophe.

If it is true, as paleontologist Giorgio Manzi (2018) suggests, that climate cycles occurring up to ten thousand years ago have continued to shape human evolution, and that from each crisis our species has selected individuals better and better suited for survival, at the same time it is insufferable to think climate crises as positive factors in species selection.

Collapsologists are the new prophets of catastrophism. Utilizing shock tactics, they exaggerate the seriousness of the environmental crisis to prepare us for a destiny of an inevitable and devastating loss. Just promulgating a survivalist discourse, they suggest navigating the collapse, asserting its absolute inevitability and the subsequent need to let be and abandon all hope, as the damage is already done and irreversible.

Nothing new: Christopher Lasch (1984) wrote that the apocalyptic vision appears in its purest form not in the contention that the nuclear arms race or uninhibited technological development might lead to the end of the world, but in the contention that a saving remnant will survive the end of the world and build a better one.

This reduction of human development trajectories to a singular catastrophic path is permeated with anti-scientific obscurantism (Bondí 2022) and reinforces splits and feelings of subjective impotence. Once convinced of the inevitability of apocalyptic scenarios, people become fatalistic and hopeless, becoming apathetic, resigned, and inclined to relinquish responsibility, and catastrophes are re-normalized, perceived as part of normal course of events.

According to Susan Sontag (1965), from a psychological point of view, the imagination of disaster does not greatly differ from one period in history to another. But from a political and moral point of view, it does. The expectation of the apocalypse may be the occasion for a radical disaffiliation from society.

Sverre Varvin (2023) speaks of "depletion," referring to a psychic process in which traumatized individuals struggle against senselessness, unpredictability, and hopelessness, gradually withdrawing both mentally and socially. Increasing numbers of couples are questioning the ethicality of having children in the face of these risks.

Slavoj Žižek (1992), stating that without the imaginary reality loses meaning, dissolves, warns against the "ecology of fear," defined as a form of compulsive activism.

The incessant barrage of information about catastrophes, particularly climate change, fails to create a supportive environment for working through anguishes and fears. The fantasy of a destroyed world reflects the radical loss of safe spaces, and the subsequent inability to attend to and care for the self and the environment. Such attitudes are counterproductive, provoking hostile reactions by fostering rigid and extreme anxiety, uncertainty leading to desperation and panic, and nihilistic complacency in the face of climate chaos and biological annihilation.

"Prospective" and "prophecy" are two different ways to push human beings to change their own perspectives and to embrace action. Their epistemic and literary styles diverge: while "prospective" emphasizes probability and science, "prophecy" may combine knowledge from science with religious-like visions and revelations (Li Vigni, Blanchard, and Tasset 2022).

The future is the result of an unpredictable and indeterminate combination of the expected and the unexpected; any attempt to impose a teleological narrative is illegitimate.

In "On Transience," Freud lyrically describes a walk with a friend and a famous young poet[1]:

> "... through a smiling countryside [...]. As regards the beauty of Nature, each time it is destroyed by winter it comes again next year, so that in relation to the length of our lives it can in fact be regarded as eternal. [...] A flower that blossoms only for a single night does not seem to us on that account less lovely."
>
> (1916, 305–306)[2]

"On Transience" suggests that the environment and affectively invested objects can be experienced in a peculiar atmosphere of loss and fear of the end. The poet is only a passive witness of a possible future destruction and certainly experiences the mourning. But he does not work through the mourning; Freud introduced an original aspect of mourning, the "anticipatory mourning" of the object, which consists in the rejection of its cathexis to avoid suffering from its loss. Anticipatory mourning is a mechanism of narcissistic defense that denies the libidinal attachment as it is being developed. In this sense, beauty is lost in advance. Freud does not accept this in any way and proposes to repair and recreate the internal and external internal world. He concludes his essay with these words:

> "When once the mourning is over, it will be found that our high opinion of the riches of civilization has lost nothing from our discovery of their fragility. We shall build up again all that war has destroyed, and perhaps on firmer ground and more lastingly than before."
>
> (Ibid., 307)

Freud can guide us into the future, urging us not to be euphorically optimistic, but confident in our and our patients' restorative and constructive capacities.

Climate terrorism, for example, can promote the nihilist and passive acceptance of damage without any hope of change, an acceptance that is a form of anxiety that can be called "eco-anxiety" and strictly related to the experience of the inevitability of the climatic catastrophe.

Another possible response is the emergence of persecutory and primitive anxieties and even the activation of psychotic paranoid defenses, a clinical situation that Henri Ey (1964) has duly defined as "the pathology of freedom."

Josè Bleger (1967) has developed the concept of ambiguity, intended as the clinical evidence of a primordial state of psychic undifferentiation, the so-called "agglutinated nucleus" of ambiguity. The mind deposits such a nucleus into the outer world through a symbiotic link or bond with depository caregivers, upon whom the individual psychically depends. That deposition provides individual with an inevitable and necessary complementary dimension implying for him/her a sense of safety and belonging. When sudden changes occur in a depositary context, such as changes in the environment, the ambiguous position can become, in such circumstances, the major defense and adapting mechanism.

Silvia Amati Sas (2020), referring to the Bleger's concept of ambiguity, speaks of the ability of "adaptation to whatsoever", affirming the existence of a basic human surface psychic capacity adaptable and standardizable to any context and circumstance, and working as a mechanism for survival in extremely traumatic experiences constituting an impregnation and obnubilation that may not be perturbing, that are not signaled by anguish and that become part of the obvious, of the flatly banal implicit that inhabits us.

Lorena Preta (2019) highlights the risk that the outside can haunt and flatten our unconscious and proposes the need for continuous work on the construction of vigilant and constructive defenses, and not passively reactive and pathological ones.

Joyce McDougall (1993) points out the problem of people she called normopathic, or hypernormal, over-adapted to life, whose normality is a deficiency that affects the whole phantasmatic life and distances the individual from him/herself.

According to Laura Ambrosiano and Eugenio Gaburri (2013), when the Ego Ideal allows itself to be bent by conformism, it ends up being inspired by the group's Ego Ideal without personal mediation, whereby the superego also degrades, its archaic and megalomanic part becomes dominant and offers the individual rigid ideologies, or a saturated worldview.

Reacting to Climate Terrorism

The task of scientists is not to announce the inevitability of catastrophe and, in this way, evacuate the catastrophist discourse, but to search the conditions for tackling problems at different levels of action, considering the strength of individual and group psychic defenses. Climate change is more than a technical issue, which calls for the reinterpretation of relationships at every fold of life.

According to William B. Cameron (1963), Albert Einstein wrote these words on the blackboard of his office at the Institute for Advanced Studies in Princeton, New Jersey (USA):

> "Not everything that counts can be counted, and not everything that can be counted counts."

Einstein's quote is significant because it stresses the role that emotional subjective features play in the context if natural sciences.

While collapsologists base their arguments on data that few would dispute, the way in which they are assembled into a dystopic end-of-the-world narrative is problematic, and the production of doomsday scenarios dismantles our belief systems, our attitudes toward the future, and our sense of good and evil on a very deep level. To announce a final global catastrophe is to ignore the real facts.

According to Sontag (1989), anything in history or nature that can be described as changing steadily can be seen as heading toward catastrophe.

It is important to avoid endorsing a closed-off vision of the future, and to unlock possible futures, proposing an ecology of desire instead of an ecology only of duty and renunciation. There are always individuals, groups, cities, or regions that take up eccentric positions, inventing alternatives, adopting divergent perspectives, and building new possibilities. There is the emergence of a multiplicity of "counter-Anthropocenes" – other possible worlds that are forged in the interstices. While they often appear as forms of "resistance," they generate other modes of perceiving the world that foster new modes of acting. We need to have the collective ability to reverse the current order of priorities and to establish the absolute value of the precautionary principle. The future remains open. Every humanist has a duty to prove the prophets of catastrophism wrong. There are countless places on this planet where people are already struggling to overcome the devastating effects of the techno-industrial hubris (Chateauraynaud and Debaz 2017).

It makes sense to shift the focus from what should not be done to what should be done, to start building and narrating places where the quality of life improves, pollution is reduced by comfortable technologies, work is more evenly distributed, social tensions ease (Cianciullo 2018).

On the other hand, uncritical praise of the natural intact world assumes, paraphrasing Voltaire's *Candide,* that everything should work for the best in the best of the possible worlds. The obsessive dramatization of environmental defense mechanisms, the continued use of alarmist tones and apocalyptic interpretations, and opposition to scientific progress, can all promote processes of magmatic fusion, substitution of the symbolization processes, and identification with a group mentality in basic assumption that lends support of hallucinosis idealizations[3] (Schinaia 2019, 2022).

Apocalyptic visions have long accompanied the environmental movement. For a long time, the rallying cry of the environmental movement has been that we must act before it is too late. Dreams of a better or utopian future have been less important as a mobilizing tool than fear of the coming catastrophe or collapse. Catastrophic fears are also engendered by conservative movements that fetishize the fusion of nativism and nature by the discourse that Joe Turner and Dan Bailey (2022) conceptualize as Ecobordering. This discourse with its racialized logics casts immigration as a threat to the local or national environment and consequently presents borders as forms of environmental protection.

In contrast, Gaia Vince (2023) argues that the climate crisis already has millions of people on the move, and that number will steadily grow higher till it breaks the political structures of the planet - unless we start now to remake those structures so they can cope, and indeed benefit, from the flow of humans that is inevitable, often necessary. This migration could become part of a good adaptation to the climate and biosphere crisis. On the other hand, it is the ancient migrations that made us what we are: as humans, we evolved by cooperating and migrating.

However, different groups in the environmental movement (i.e., Greenpeace and WWF) have strongly contested the retrograde and reactionary positions of ecobordering in various ways, arguing that messages emphasizing the hope of a possible, better future should replace images of the apocalypse because the latter have no mobilizing function (Cassegård and Thörn 2018).

John Steiner (2018) explores the timeless dimension of idealized mental retreats by appealing to the so-called "Garden of Eden Illusion," where time stands still, everything is perfect, and nothing can change; there is no development, frustration, loss, or passion. If the trauma of disillusion occurs too abruptly or the suffering of waiting is too intense, defenses are mounted against the impact of reality. These defenses lead to a misrepresentation of the reality of life in which experiences involving time are refused in favor of instant solutions based on omnipotent thought. Loss and conflict spawn defensive movements such as denial, distancing dynamics, apathy, and eventually narrow-mindedness.

The Working Hope

> We must accept finite disappointment but never lose infinite hope.
> Dr. Martin Luther King in Washington, DC, Address in February 1968

Many emerging phenomena experienced as cataclysmic penetrate the core of mental experience and can reshape the unconscious interactions among human beings in new and often dramatic ways. However, we must maintain a distance from oppressive cataclysmic experiences and shape our reflections by embracing the condition of wandering – constantly questioning an elusive, versatile, plural, and ambivalent reality. Individually and collectively, we must inhabit it with discretion, doubtfulness, flexibility, and caution.

We must carefully observe lifestyle modifications that are as intense as they are confusing. They can be vectors of change that swiftly alter their trajectory in emergency situations. It is crucial to recognize the right value of signaling social behaviors, but we must not see their significance as absolute, avoiding drawing catastrophic "everything has changed" conclusions or, conversely, consolatory "essentially nothing has changed" conclusions.

Ulrich Beck (2016) proposes the concept of "Metamorphosis" that implies a radical transformation in which the old certainties are falling away, and something quite new is emerging.

Every genuine progression challenges our tolerance for the uncertainty of the "truths in transit," (Horovitz 2007) truths that move us away from the risk of thinking we always have the final and definitive solution at hand. They are small truths, probably slightly larger than a babble expressing a desire. However, we cannot exclude these truths because they support and promote psychic transformations. They can be usefully and deeply explored provided they do not lose their connotation, that is, that status of truths in transit towards the construction of new identities, which necessarily require time and patience. Transitory identities, propaedeutic perhaps to future stable identities, but authentic and authenticated by relational mutual consent. These identities are thought of as to be constructed in the bumpy paths of the analytical relationship with the materials at hand, rather than original entities to be discovered. This means that we can open up and experiment new cognitive and relational paths, even if we start from familiar landscapes.

In the exergue of the last chapter of his *Attention and Interpretation* (1970, 125), Bion refers to the "Negative Capacity" by referring to a part of the letter the English poet John Keats sent to his brothers George and Thomas of December 21, 1817 on "what quality went to form a Man of Achievement." The "Negative Capacity" is "when a man is capable of being in uncertainties, mysteries, doubts, without any irritable reaching after fact and reason."

This Negative Capacity permits toleration of differences, changes in points of views, the uncertainties in the searching of the adequate preventive and therapeutic solutions. It is a capacity that permits us to be ourselves and to keep alive the desire to comprehend and learn without the need to fill the void at any cost.

The words of Keats and the reflections of Bion propose dealing with the vicissitudes of existence by accepting the uncertainty and the complexity, by avoiding the anti-economical illusion of thinking to be able to control and manage what frequently is uncontrollable and unmanageable. If we become facile in our attempts to reduce something unknown to something known, what is incongruous to what is congruous, we run the risk of becoming partners in crime, engendering resistances related to anxiety and detaching from a non-immediate solution of the problems.

As human beings and psychoanalysts, we must invent/find new realms of humanity – places where intimacy, thoughtfulness, and pauses are revered. These spaces allow for face-to-face encounters that foster new ideas and bring forth small utopias, perspectives of meaning that continually evolve rather than fixate on a definitive destination. They are itinerant utopias, capable of generating in us the capacity to pause, to ask how the journey is going, what is happening on the way (Ambrosiano 2017).

Ernst Bloch (1954–1959) has distinguished between abstract hope and concrete hope. Abstract hope pertains to the realm of wishes. This version of hope propagates a romantic or optimistic view that fosters a rather passive position, one of trusting that the future will fatalistically bring change. Contrary to abstract hope, hope as "concrete utopia" is closer to the present in its participation and awareness of current obstacles. What interests Bloch is not so much what is or what has been, but rather the "ontology of the not-yet-being" toward the emergence of novel possibilities that are always already latent in the world, the "figures of hope" with fore- shadow the human potential.

While abstract hope may be viewed as manic defense, concrete hope embraces the limitations one experiences in the now (Tylim 2019).

According to Rebecca Solnit (2004), hope is not a sunny everything-is-getting-better narrative, although it may be a counter to the everything-is-getting-worse one. Hope can be called an account of complexities and uncertainties, with openings. Critical thinking without hope is cynicism, but hope without critical thinking is naivety.

Clive Hamilton (2010) maintains that, before undertaking anything at all, we have to purge hope from our desperately optimistic framing of life.

Bruno Latour (2015) says that, instead of speaking of hope, we would have to explore a rather subtle way of "dishoping"; this doesn't mean "despairing" but, rather, not trusting in hope alone as a way of engaging with passing time.

In *The Book of Hope* (Goodall, Abrams, and Hudson 2021), naturalist and primatologist Jane Goodall asks: How do we stay hopeful when everything seems hopeless? How do we cultivate hope in our children? What is the relationship between hope and action?

Then, she focuses on her "Four Reasons for Hope":

1   The Amazing Human Intellect. While an intelligent animal would not destroy its only home as our species is doing, humans have the intellectual power to come up with new innovations all the time, including renewable energy, regenerative farming and permaculture, moving toward a plant-based diet.
2   The Resilience of Nature attested to by the example of Haller Park in Kenia, where injured Earth has been restored and healed.
3   The Power of Young People, from elementary school age right through to college. It's crucial that young people know how positive action can still shift the frightening trajectories of climate crisis and biodiversity loss.
4   The Indomitable Human Spirit, the ability humans have individually and collectively to wrest a victory from what appears to be an inevitable defeat.

Short Notes on Hope in Psychoanalysis

In psychoanalytic theory we can trace two different approaches to hope, in line with Bloch's distinction. The first approach views hope as essentially regressive and an obstacle to a mature and rewarding experience, the second views it is essentially progressive and conducive to a richer experience. Stephen Mitchell (1993) distances himself as much from a purely regressive hope as from a purely progressive hope, emphasizing its ambiguity, and testing the possibility of new growth grafted onto old hopes. He proposes hope as the dialectical relationship between what is static and familiar, and the desire for something richer and more rewarding. We require a mature and depressive working hope that skirts the childish notion of "all and now" and instead fosters firm and enduring the anti-entropic efforts against forces pushing us towards the dissolution of bonds. Although Erik Erikson (1968) considers the origin of hope in childhood, he does not give it a regressive meaning, but a constructive one in the later stages of development, in the sense that it promotes growth.

Eugenio Borgna (2020), echoing Kierkegaard's words, defines this hope as the "passion of possible," a beacon of light that does not let us be submerged by darkness, because it knows how to grasp the lights of the future. It contrasts the nihilistic and passive acceptance of damage without any hope of change. Increasingly, there is a prevalence of desperately destructive attitudes such as "there's nothing to do now," "you have to live for the day," "politics sucks," etc., which are malignantly narcissistic attitudes undermining the legacy of the new generations. We must provide our patients with words that value small changes, small novelties, as antidotes against poisonous idealization-ideologization that pushes towards the unattainable and makes them experience the achieved as taken for granted and disappointing. They are often imbued with this poison or, on the contrary, with a disappointed, muffled, and frustrating resignation, which frequently pushes them towards the loss of memory of the past, the trivialization of the present, and the nihilization of the future, in a sort of alexithymic existence.

Earl Hopper (2006) suggests that in mature, authentic hope, despite obstacles and adversity, or perhaps because of them, a certain fortitude, optimistic thrust of life, and tenacity will prevail over feelings of despair and bitterness. The development and the maintenance of a hopeful attitude are characteristics of a mature person who is able and willing to take the role of "citizen". A mature citizen is one who is willing and able to try to create within the context of life trajectories the social, cultural, and political conditions that are necessary for the development and maintenance of a hope finite and contingent – for the organization of hope, as Pope Francis says for others as well as for oneself.

Franco Fornari (1985) views psychoanalysis as the revival of hope capable of mobilizing the basic trust that is part of our vital endowment.

In her work with traumatized patients, Alessandra Lemma (2004, 125) defines mature hope as "a state of mind of expectant possibility routed in a profoundly moving and sobering appreciation of the 'possible', or, if you like of reality." Lemma links the development of mature hope to the integration of one's own potential for love and for hate.

Isaac Tylim (2007) says that the privacy of the consulting room has the potential to offer a unique opportunity to co-construct a safe transitional space where hope could be restored. Waiting might be viewed as a container that embraces disruptive affects and as a means of mobilizing agency. Without waiting, hope gets diluted in despair. When hope fails and the dangers of the present threaten the therapeutic dyad with despair, turning to the utopian imaginary may offer a lifeboat to navigate turbulent waters. Utopia helps to survive the impossible present while charting the course for a new and different future via hope of the concrete variety.

Steven H. Cooper (2014) sees as a central dialectic tension in psychoanalysis: that between psychic possibility and psychic limit. It is only by accepting the realm of limit as a necessary counterpoise to the realm of possibility, and clinically embracing the tension between the two realms that analysts can further their understanding of therapeutic process.

Giuseppe Pellizzari (2015) describes hope as oscillating between the passivity of confident expectation, and the activity of a passionate search. Distinct from certainty, and born of anxiety and doubt hope, pointing a perspective and activating potentialities directed towards the unknown, is a powerful transformative factor. It does not have an object, it is a utopian tension, and it is unsaturated; it disturbs the fundamentalist certainties of the unconscious that are subject to the repetition compulsion that imprisons the perception of reality in a fixed and dogmatic scheme, paralyzing experience and knowledge.

Maurizio Balsamo (2023) sees hope as the articulation between the field of desire and the field of the negative that opposes it. For him, the hope principle, in the analytic treatment, is given by the possibility of accessing the functions of assembling/disassembling of scenes, memories, constructs, personal theories, in order to reopen the field of the possible, the area of play in the timelessness of fate.

In *On Giving Up* (2024), Adam Phillips paradoxically illuminates both the gaps and the connections between the many ways of giving up that is as important to our psychological well-being as hope and love. For him renunciation is the fulcrum of the change. To give up on hope would just mean to give up on wanting it, just as giving up is always giving up wanting something or someone.

Radical Hope

Jonathan Lear's concept of Radical hope (2008) offers a visionary approach to navigating turmoil by redefining ways of existence amid cultural upheaval. Lear reflects on how Crow Chief Plenty Coups led his people through a time of great turmoil. Faced with the destruction of the Crow's cultural ways of living, when they were forced to live on reservations and give up buffalo hunting, he was able to propose imaginative reshaping, leading his people to redefine their very ways of existing, so that they would not fall into utter despair. What makes the hope "radical" is that it is directed toward a future goodness, anticipating a good for which those who have the hope as yet lack the appropriate concepts with which to understand it. The hope is "radical," because it is virtually impossible to say beforehand what the shape of this new kind of life will be. This has to emerge from disquiet and doubt in specific new forms, drawing on the particular cultural resources of each society. Lear delves into what he calls the "blind spot" of every culture: the inability to conceive of its own devastation and to propose possible original solutions, openings to the inconceivable, as is the case in these times when natural disasters leave us with "an uncanny sense of menace," and a heightened perception of how vulnerable our civilizations are to destruction. Radical hope can help us not to fall into despair or accept the *status quo*. More recently, Lear (2022) considers our bewilderment in the face of planetary catastrophe and asks how we can live with the knowledge that cultures, to which we traditionally turn for solace, are themselves vulnerable. Lear argues that we need to repeatedly engage with the past and explores how creative mourning can be an alternative to despair and help us in our development, and what place gratitude and care should take in human life.

Václav Havel (1991) says that hope is not the conviction that something will turn out well, but the certainty that something makes sense, regardless of how it turns out.

Pope Francis similarly in the encyclical "Laudato Si'" (2015) and in the apostolic exhortation "Laudate Deum" (2023) advocates a realistically pessimistic but not despairing attitude, in a sense close to that of Antonio Gramsci[4], who advocates pessimism of the intellect and optimism of the will. He states that hope consists in recognizing that there is always a way out, that we can always change course, that we can always do something to solve problems, despite the great speed of changes and degradation.

Developing the New Bioethics for the Future

The current crisis caused by pollution and biodiversity loss have made us all deeply aware of our planetary precarity, so we look for ways to adjust, recalibrate and mitigate, and remain stubbornly positive and relevant. It is not easy to accept the constant shifts within nature and its transience. This difficulty meets/clashes with the desire and economic need to use common goods for one's own needs that in fact greatly damage our future and the future of our children and grandchildren.

Our current challenge is to create and develop the new bioethics for the future that encompasses the complexity of global reality. When Aldo Leopold (1949) proposes "thinking like a mountain" land ethic, stresses that the relationships between people and land are deeply intertwined: care for people cannot be separated from care for the land. Land ethic is a moral code of conduct that grows out of these interconnected caring relationships.

Archaeology and history have investigated the socio-political, economic and environmental reasons that have led complex societies such as Egyptians, Romans, Mayas, Incas, Minoans, or Mesopotamians, to disappear or decline. Several concurrent or complementary explanations have been given by scholars and experts: environmental modifications, techno-economic decreasing returns, and elite incompetency are just some examples. Interestingly, most of these works have tried to learn lessons from the past in order to warn present societies against the risks of collapse (Li Vigni, Blanchard, and Tasset 2022).

Jared Diamond (2005) accurately avoids any appeal to catastrophism and argues that ancient civilizations were able to re-examine the relationship with the Earth as a system. They successfully faced environmental challenges and survived disasters.

Loss and conflict often trigger defensive movements such as denial, distancing, impotent rage, or apathy, and hinder transition and action. Group anxieties may take forms of basic assumption (Bion 1952).

A strategy to counter them involves giving people tools to name and explore together their most difficult emotions to sit with painful feelings without instantly running away from them, to make them emotionally secure in order to create areas of dialogue and collaboration to reduce the intensity of their defenses, share their internal worlds, and connect with their creativity and capacity to mend.

Creating conditions for participation, avoiding drastic judgment, reducing the space of the "environmental Super-Ego," fostering collaboration through finding the "idiorrhythmy," the "right rhythm," a productive form of living together in a community wherein everyone is able to retain their own rhythm and recognize and respect the individual rhythms of the others (Barthes 2002). All these are means to contain and regulate the effects of the environmental crisis, encourage people to explore their interior dilemmas, and promote solicitude, creativity, and solidarity in the face of our vulnerability to disasters. Uncertainty can thus become an important driver of growth and development, more an opportunity than a threat.

Like Odysseus, we must navigate between the Scylla of the mere negationist optimism or the regressive hope, and the Charybdis of the nihilist pessimism and fatalistic inevitability where recognizing the problem is strictly connected to the desperate impossibility of immediately solving it. It is impossible to respond to the difficulties proposed by the environmental crisis with the manic hope of repair through easy solutions – a hope often followed by disillusionment and the end of expectations if the solution does not arrive rapidly, within the desired timeframe.

Proposing a planetary humanism, Mauro Ceruti and Francesco Bellusci (2023) say that we are facing the end of a world, that is the catastrophe of Modernity as it has been conceived so far, rather than the end of the world.

There is a crucial difference between survivalist fantasies of the end of civilization and those who speak in terms of the end of civilization as we know it. Whereas the former is consumed by anticipation of the end of the world, the latter is glimpsing the possibility of new beginnings (Hoggett 2023).

Virginia Ungar (2015) invites the analysts to reflect upon the form and the use of the psychoanalytic tools, because we are contemporaries of a series of socio-cultural changes and, as consequence, of transformations in subjectivity. To face a new reality, we must use both new and old tools and go beyond what we knew, as Mark Halle proposes about a new Nature narrative against catastrophism. We don't need policies that blame and give space to an apocalyptic terror regarding environmental change, or simply propose a list of good and right practices for saving the planet without considering how emotional difficulties hinder people from understanding and putting these practices into action.

Jonathan Franzen (2019) suggests that we can solve the dilemma between trying to avoid the inevitable climatic catastrophe and preparing to defend us against its negative effects by renouncing to invert the effects of global warming. We must focus on more easily achievable short-term goals through immediately feasible collective actions. These actions can allow us to have one less hurricane and a few more years of stability.

The only answer that can counter the impotence, the inertia, and the apathy that the present environmental disaster provokes in all of us is to try to find and experiment with possible common remedies. This requires us to conceive humankind simultaneously as a collective and as single individuals.

In a time marked by impending catastrophes, a politics able to face desperation, loss, emotional paralysis, and marginalization, and to propose free and creative projects for the future is far more useful than a politics that blames, gives space to the apocalyptic terror. We need a politics of solidarity and opposition to the idea that the human condition is an immutable destiny. The precarity of hope should be preferred to the certainty of despair.

According to Varvin (2023), psychoanalysis can represent a mediating function in relation to regressive tendencies at individual, group and societal levels. He introduces a conceptualization of a third position in psychoanalytic clinical work where anxieties can be contained, understood and reflected upon, emphasizing the necessity of anchoring symbolization and working through in a common cultural discourse. This model is not a neutral or theoretical stance but a need for a certain flexible distance from which to observe, think and engage. It may lay the groundwork for functioning as antagonistic way in mass-regressive situations and understanding how social catastrophes can be worked through at the individual and social levels.

Adrienne Rich (2003) says that despair, when not the response to absolute physical and moral defeat, is, like war, the failure of imagination.

Through dreaming, thinking and rethinking what happens inside and outside of us, we can structure various and original forms of language and experience. These forms cannot be the sum of the previous languages and experiences but something emerging from them, with their own configuration and autonomous and original life. Common remedies are possible by building containers in which there is a creative conversation among different scientific and cultural languages, as the authors of this book try to do. This conversation tries to connect different and disconnected areas of knowledge. It avoids can take narcissistic individualism and pretentious conquering ambition, without necessarily seeking totalizing harmony but helping us become hospitable to and accommodating of other people's reasoning and feelings. A more evolved and elaborate use of one's own ideological, scientific, cultural, or personal beliefs can lead to unsaturated positions and to reparative tendencies. In these tendencies, concern and responsibility for the life and destiny of the individual and the community predominate.

New ideas and projects need considerable time to be realized. Especially at the beginning, they are often harshly rebuffed. Winnicott (1971) argues that one of the main tasks of mothers is to propose a new object to their babies. This task is a real art. On the one hand, if a mother forces the issue and pushes the baby to approach the new object too much, she will receive a rejection as a response. On the other hand, if she proposes a "period of hesitation" during which the baby will probably turn away and not show any interest, she will see that the baby shows a great interest in the new object.

Conclusion

To avoid removing ecological catastrophes caused by development without rules or memory – what could be called "cannibalistic development," we cannot collude with the culture of decline, but we need to look deeper into individual, family and social dynamics and conflicts.

According to Diamond (2019), by resorting to a process of painful self-analysis and looking to the future, individuals and groups can address current crises through a process of selective change. He suggests selectively assessing aspects of individuality that function well and should not be modified and those requiring change to unlock new solutions tailored to each person's capacities.

In his posthumous notes, De Martino (1977) uses the thought-provoking term: "vital catastrophe," while Bion (1965) discusses catastrophic change as a basis for vital transformation.

Recently, three books have challenged, albeit perhaps too optimistically, the catastrophist ways in which the environment is talked about suffering from unreasonable prejudices.

In *Factfulness* (2018), Hans Rosling and his colleagues suggest that the vast majority of people are wrong about the state of the world because the media systematically skew data and trends and select stories to make people think that the world is getting worse. According to them, we are facing a radical improvement and we must not allow ourselves to be blinded by anger, ignorance and oversimplification, but must instead learn to look at the facts with curiosity, to put them into perspective and to know how to amaze ourselves: we need only think back to the lives of our grandparents to realize the enormous strides we are making, in every field.

In his *Rien n'est Joué* (2023), Jacques Lecomte, a proponent of ecological opti-realism, an optimism aware of the extent of the difficulties facing the world, argues that the collapse advocated by the collapsologists is scientifically false and generates an existential angst that may even justify in the eyes of some people a return to authoritarian policies.

In her *Not the End of the World* (2024), a bold, radically hopeful book, data scientist Hannah Ritchie argues that we must rethink almost everything about the environment, dismantling the narrative of the good old world, which was never good, just and sustainable. The data shows we've made so much progress, and so fast that, far from being the last generation to live on Earth, today's young people could be the first generation on track to achieve true sustainability in history. We have the real opportunity to be the first generation to build a sustainable planet, leaving the environment in a much better state than they found it. She is no climate change denier, or minimizer; on the contrary she brings attention to the magnitude of its potential impacts but stresses that we are on a better trajectory than most people think.

Of course, it must be carefully avoided that denialism disguised as anti-catastrophism will lower the guard with respect to the need to cope with the climate crisis now.

There are authors, such as Bjørn Lomborg, whose books, the first *The Skeptical Environmentalist* (2001) and the last *False Alarm* (2021), who can be called "lukewarmers". Lomborg argues that climate change is a real problem caused by humans, but a manageable one, and that it will not destroy the world, but rather slightly slow its progress. His fraudulent cherry picking which consists of ignoring all the environmental data that might refute his theses and highlighting only the evidence in his favor, is strongly criticized by the scientific community.

Moving away from splitting phenomena toward greater integration is neither simple or free of danger, demanding introspection devoid of complacency – a journey fraught with complexity and risks. Establishing social containers capable of transformative containment is essential for fostering care and empowerment at both individual and group levels (Di Chiara 1999).

Just as psychoanalysis needs ecology, ecology needs psychoanalysis to develop and bolster Ego resources for mature engagement with new existential challenges and to meet contemporary individual and social emotional difficulties with critical understanding and a creative approach to catastrophes in our time.

Mariano Horenstein (2024) proposes psychoanalysis as a critical thinking, a space for undoing the normalization of catastrophe in the West, the anesthesia induced by repeated exposure to horror images, ongoing wars, and a world order marked by gross inequality in resource distribution. He suggests that perhaps one of humanity's last strongholds lies in remaining perplexed, in not losing the capacity to feel fear, and in transforming noises into voices, voices into words, words into stories worth hearing.

We can counter catastrophism with the working hope in the face of contemporary reductionism, which seeks to erase doubt and heterogeneity, and also valorize hope in the difficult situations, as Winnicott (1967) says about delinquency. According to him, the antisocial tendency, intrinsically linked to child deprivation, begins to turn into real delinquency in the boy or girl, if the hoped-for communication does not develop because the act is either ignored and not recognized as a signal of hope, as something that contains an SOS, or is read as solely negative rather than partly positive.

The future remains open and the catastrophe we witness is not solely the representation of a tragic end but also the embodiment of a formidable movement and work of thought.

Paul-Laurent Assoun (2023) proposes an investigation of the unconscious aspects of the catastrophic phenomenon to arrive at the transition from the Freudian discontents of civilization to the catastrophe in civilization. The long time of reflection and the contracted time of catastrophe are indeed "dis-cronic", but to "think catastrophe" is to attempt to re-found its own temporality, that of "rupture" and heterogeneity. In this way, catastrophe is the scenario of the unforeseeable, the entry into the extraordinary, and thus the way out of the "common". It turns out to be a source of exceptional awakening power, beyond the sense of dismay.

A Paradise Built in Hell (2009) by Rebecca Solnit is an eye-opening account of how much hope and solidarity emerge in the aftermath of disaster. She argues that disasters often produce remarkable temporary communities – paradises of a sort amid the rubble, where human beings, acting on their own and without direction from the authorities, manage to provide for each other. Disasters can be opportunities to rediscover the powerful engagement and purposeful joy of altruism and generosity.

At the end of *Invisible Cities* (Calvino 1972), Marco Polo says that the hell of the living beings is not something that will occur in the future. If there is a hell, it is what is already here, what we construe by being together. There are two ways to escape it. The first is easy for many: accepting the hell and becoming such a part of it that you can no longer feel or perceive it. The second way is a risky one that requires constant vigilance and apprehension: it requires finding out who and what in this hell are not parts of the hell and making them endure and giving them space.

Finding out what is not hell and giving it space is an exhortation to heal the wounds that we inflicted on the planet, maintain the places, repair the damages of time and human being.

We must avoid denials and catastrophism, those psychic fossilizations, which are widely present in us and transmitted to the next generation. We cannot reproduce our planet. The damages we wrought are a sort of bank-rupting of our descendants. Generational transmission implies an identification process that summarizes a history that, for the most part, does not belong to future generations. When we underestimate the contract that links everyone to the whole and vice versa – us to the Earth and the Earth to us – we necessarily transmit symptoms, defense mechanisms, organization in the object relationships, the signifiers, all the elements in which the forms and the processes of the psychic reality of a single subject articulated with the forms and the processes that constitute in the intersubjective relationships and, more in general, in our relationship with our environment (Kaës 1993).

Recognizing these dynamics is the starting point to changing individual and family dynamics and lifestyles to allow, in a rediscovered dimension of fraternal collaboration, sustainable action that is creative, respectfully restorative and potentially part of a global renewal, a reassumption of individual responsibility that, in turn, can stimulate collective responsibility. We must construct a horizon of meaning that rigorously refers to the principle of reality and opposes the skepticism of those who think that the individual is condemned to impotence, trapped in a sort of suicidal environmental melancholia. We must rediscover the pleasure of personal responsibility, of taking care of ourselves condition and of those who depend on us as an antidote against fear and indifference. This pleasure and the capacity of re-establishing the contact with the emotional experience of trust in a caregiver can allow us to uncover unknown energies and use them to secure ourselves and other people (Ferruta 2020).

The environment is a collective good, the heritage of all humanity, and the responsibility lies with everyone; no one is excluded. Edgar Morin (2011) points to the horizon of a new planetary humanism, in which we can recognize ourselves as a community of destiny in which all individuals and all peoples are united in humanity and humanity itself in the ecosystem of its Homeland Earth.

The crucial task of psychoanalysts is to thoroughly investigate the new psycho-social realities, the new discomforts of civilization, the new forms of mental suffering linked to the crumbling of traditional identity structures and the inadequacy of the metapsychic guarantors, the concepts that defined reality in the past. By valuing the ability to think and dream, we can offer a counter to the feelings of disaster, the sense of an end of the world, that increasingly grip people.

To face a new reality, we must use both new and old tools and go beyond what we known. Thus, Pierre Fédida (2007) suggests that the analyst's role is to imagine what another person has experienced.

We must risk new analyses, create new mental tools, propose new modes of comprehension, allowing us to consider again and provisionally the relationship with the stranger we chose as our way of being in the world (Kaës 2012).

Jonas (1979) opposes the principle of despair of Günther Anders (Portinaro 2003), according to whose rhetoric of intransigence, it is no longer permissible to even hope because he considers hope as the renunciation of action the human condition to be essentially irretrievable. Jonas argues that, according to the principle of responsibility in the light of the ecological crisis, we must have a meaningful perspective that goes beyond our single lives. Our successors have the right to expect a world that is in at least not in worse condition than we found it. His utopian ecological imperative is: "Act so that the effects of your action are compatible with the permanence of genuine human life." (Chapter 1/V) Or phrased the other way round: "Act not destructively for future generations and the totality of their life conditions."

While knowing that the work in the consulting room is a small and albeit difficult thing in the face of organized human living in groups and societies, we can envisage a good enough life, in which there can be room for love and creativity, counteracting magical and illusory thinking, and looking with sincerity and integrity even at the negative aspects of existence, offering the means to experience them with reflective consciousness through the patient and continuous symbolization work the patient and continuous work of symbolization that counteracts the power and rigidity of inner defensive mechanisms, from which none of us individuals are immune.

Here a few sentences from the speech Gabriel Garcia Márquez gave when he was awarded the Nobel Prize in 1982:

> "To oppression, plundering and abandonment, we respond with life. Neither floods nor plagues, famines nor cataclysms, nor even the eternal wars of century upon century, have been able to subdue the persistent advantage of life over death. [...] We [must] feel entitled to believe that it is not yet too late to engage in the creation of a new and sweeping utopia of life, where no one will be able to decide for others how they die, where love will prove true and happiness be possible, and where the races condemned to one hundred years of solitude will have, at last and forever, a second opportunity on earth."

We can find a way to reinhabit the world in all its multiplicity, ethically and sensitively caring for its wounds and ruins. In this world we inhabit, each living organism impacts the conditions of life for other organisms.

Notes

1 Freud's friends are probably Lou Andreas Salomé and Rainer Maria Rilke.
2 Bion too (1990, 100) focuses on the emotion that the primitive part of the mind registers upon confronting fundamental reality. He also stresses the risk of apathetic intellectualization emanating from the scientist reductionism:
   "[...] I would like to be capable of being awe-inspired by a sight like the aurora borealis. I would not think that I had improved if I said, 'Oh well, this is simply an electric display; it's an electric phenomenon.' That is clever, but not wise."

3 Bion (1965) describes transformation in hallucinosis (TH), as a psychic defense present in elusive psychotic scenarios in which there is a total adherence to concrete reality.

4 Pessimism of the intellect, optimism of the will" is a motto borrowed from the French writer Romain Rolland. Antonio Gramsci expanded and developed this initial phrase, giving it a new and richer meaning. He explicitly attributes it to Rolland, who used it a few days before, in the review of *The Sacrifice of Abraham* by Raymond Lefebvre (Paris: Flammarion, 1919). The review appeared in *L'Humanité*, 19 March 1920.

References

Amati Sas, S. (2020). *Ambiguità, Conformismo e Adattamento alla Violenza Sociale.* Milan: FrancoAngeli.

Ambrosiano, L. (2017). "Fuori". Speech at CMP (Psychoanalytic Centre of Milan), January 19. Unpublished.

Ambrosiano, L. & Gaburri, E. (2013). *Pensare con Freud.* Milan: Cortina.

Assoun, P.L. (2023). *Psychanalyse de la Catastrophe: Enjeux Anthropologiques et Cliniques.* Paris: PUF.

Balsamo, M. (2023). "Nachleben. Il principio speranza nella cura analitica". *Knot-Garden*, 4: 44–64.

Barthes, R. (2002). *How to Live Together: Novelistic Simulations of Some Everyday Spaces.* K. Briggs (Trans). New York: Columbia University Press, 2012.

Beck, U. (2016). *The Metamorphosis of the World: How Climate Change is Transforming Our Concept of the World.* Cambridge, UK: Polity Press.

Bion, W.R. (1952). "Group dynamics: A re-view". *International Journal of Psycho-Analysis* 33(2): 235–247.

Bion, W.R. (1965). *Transformations: Change from Learning to Growth.* London: Heinemann, 2013.

Bion, W.R. (1970). *Attention and Interpretation.* London & New York: Routledge, 1984.

Bion, W.R. (1990). *Brazilian Lectures: 1973 São Paulo; 1974 Rio de Janeiro/São Paulo.* London & New York: Routledge.

Bleger, J. (1967). *Symbiosis and Ambiguity: A Psychoanalytic Study.* J. Churcher & L. Bleger (Eds). London & New York: Routledge, 2012.

Bloch, E. (1954–1959). *The Principle of Hope.* N. Plaice, S. Plaice & P. Knight (Trans). Cambridge, MA: MIT Press, 1986.

Bondí, R. (2022). "Catastrofi e catastrofismi". *Psiche, Rischio*, 9(2): 599–610.

Borgna, E. (2020). *Speranza e Disperazione.* Turin: Einaudi.

Brecht, B. (1939). *Mother Courage and Her Children.* R. Manheim (Trans). New York: Random House/Pantheon Books.

Byron, G.G. (1816). "Darkness". In S. Greenblatt (Ed), *The Norton Anthology of English Literature.* Vol. D (8th ed.) (pp. 614–616). New York: W.W. Norton & Co., 2006.

Calvino, I. (1972). *Invisible Cities.* W. Weaver (Trans). San Diego, CA: Harcourt, 1974.

Cameron, W.B. (1963). *Informal Sociology: A Casual Introduction to Sociological Thinking.* New York: Random House.

Cassegård, C. & Thörn, H. (2018). "Toward a postapocalyptic environmentalism? Responses to loss and visions of the future in climate activism". *Environment and Planning E: Nature and Space*, 1(4): 561–578. doi:10.1177/2514848618793331.

Ceruti, M. & Bellusci, F. (2023). "Oltre il catastrofismo. Umanizzare la modernità". *Psiche, Finimondo*, 2: 381–391.

Chateauraynaud, F. & Debaz, J. (2017). *Aux Bords de l'Irréversible. Sociologie Pragmatique des Transformations*. Paris: Petra.

Cianciullo, A. (2018). *Ecologia del Desiderio. Curare il Pianeta senza Rinunce*. Sansepolcro: Aboca.

Cooper, S.H. (2014). *Objects of Hope: Exploring Possibility and Limit in Psychoanalysis*. London & New York: Routledge.

De Martino, E. (1977). *The End of the World: Cultural Apocalypse and Transcendence*. D.L. Zinn (Trans). Chicago: The University of Chicago Press, 2023.

Di Chiara, G. (1999). *Sindromi Psicosociali. La Psicoanalisi e le Patologie Sociali*. Milan: Cortina.

Diamond, J. (2005). *Collapse: How Societies Choose to Fail or Succeed*. London: Penguin Books.

Diamond, J. (2019). *Upheaval: Turning Points for Nations in Crisis*. Boston, MA: Little, Brown and Company.

Dupuy, J.-P. (2002). *How to Think About Catastrophe: Toward a Theory of Enlightened Doomsaying*. M. DeBevoise & M. Anspach (Trans). East Lansing, MI: Michigan State University Press, 2023.

Erikson, E.H. (1968). *Identity: Youth and Crisis*. New York: Norton.

Ey, H. (1964). "L'Essence de la maladie mentale et la loi de 1838 (Aliénation, espace et liberté)". *L'Évolution Psychiatrique*, 29(1): 1–5.

Fédida, P. et al. (2007). *Humain/Déshumain. La Parole de l'Œuvre*. J. André (Ed). Paris: PUF.

Ferruta, A. (2020). "Coronavirus: a sphinx of modern times". Retrieved from www.cmp-spiweb.it/coronavirus-a-sphinx-of-modern-times.

Fornari, F. (1985). *Affetti e Cancro*. Milan: Cortina.

Franzen, J. (2019). *What if We Stopped Pretending?* London: Fourth Estate, 2021.

Freud, S. (1916). On Transience. *S.E.* 14: 303–307.

Goodall, J., Abrams, D., & Hudson, G. (2021). *The Book of Hope: A Survival Guide for Trying Times*. New York: Celadon Books.

Hamilton, C. (2010). *Requiem for a Species: Why We Resist the Truth about Climate Change*. London & New York: Routledge, 2015.

Havel, V. (1991). *Disturbing the Peace: A Conversation with Karel Huizdala*. New York: Vintage.

Hoggett, P. (2023). *Paradise Lost? The Climate Crisis and the Human Condition*. Simplicity Institute Publishing.

Hopper, E. (2006). "On the nature of hope in psychoanalysis and group analysis". *British Journal of Psychotherapy*, 18(2): 205–226.

Horenstein, M. (2024). "Freud a Gaza. Lo psicoanalista come testimone uditivo". In M. Francesconi & D. Scotto di Fasano (Eds), *Freud a Gaza* (pp. 53–85). Pistoia: Petite Plaisance.

Horovitz, M. (2007). "Transfert et vérité". In F. Guignard. & Th. Bokanowski (Eds), *Actualité de la Pensée de Bion* (pp. 44–51). Paris: In Press.

Jonas, H. (1979). *The Imperative of Responsibility: In Search of an Ethics for the Technological Age*. H. Jonas & D. Herr (Trans). Chicago, IL: University of Chicago Press.

Kaës, R. (1993). "Le Sujet de l'Héritage". In R. Kaës, H. Faimberg, M. Enriquez, & J. J. Baranes, *Transmission de la Vie Psychique entre Générations* (pp. 1–16). Paris: Dunod.

Kaës, R. (2012). *Le Malêtre*. Paris: Dunod.

Klein, N. (2007). *The Shock Doctrine: The Rise of Disaster Capitalism*. London: Penguin.

Lasch, C. (1984). *The Minimal Self: Psychic Survival in Troubled Times*. New York: W.W. Norton & Co.

Latouche, S. (2015a). "Disaster, Pedagogy of". In G. D'Alisa, F. Demaria, & G. Kallis (Eds), *Degrowth: A Vocabulary for a New Era* (pp. 94–96). London & New York: Routledge.

Latouche, S. (2015b). "Decolonization of imaginary". In G. D'Alisa, F. Demaria, & G. Kallis (Eds), *Degrowth: A Vocabulary for a New Era* (pp. 117–120). London & New York: Routledge.

Latour, B. (2015). *Facing Gaia: Eight Lectures on the New Climatic Regime*. C. Porter (Trans). Cambridge, UK: Polity Press, 2017.

Lear, J. (2008). *Radical Hope: Ethics in the Face of Cultural Devastation*. Cambridge, MA: Harvard University Press.

Lear, J. (2022). *Imagining the End: Mourning and Ethical Life*. Cambridge, MA & London: Belknap, Harvard University Press.

Lecomte, J. (2023). *Rien n'est Joué. La Science Contre les Théories de l'Effondrement*. Paris: Les Arènes.

Lemma, A. (2004). "On hope's tightrope: Reflections on the capacity for hope". In S. Levi & A. Lemma (Eds), *The Perversion of Loss: Psychoanalytic Perspectives on Trauma* (pp. 108–126). London & New York: Routledge.

Leopold, A. (1949). *A Sand County Almanac*. New York: Oxford University Press.

Levine, D.P. & Bowker, M.H. (2019). *The Destroyed World and the Guilty Self: A Psychoanalytic Study of Culture and Politics*. Bicester: Phoenix.

Li Vigni, F., Blanchard, E., & Tasset, C. (2022). "Theories of global collapse: Closing down or opening up the futures?". *Journal of Future Studies*, 27(1): 1–14.

Lomborg, B. (2001). *The Skeptical Environmentalist: Measuring the Real State of the World*. Cambridge, UK: Cambridge University Press.

Lomborg, B. (2021). *False Alarm: How Climate Change Panic Costs Us Trillions, Hurts the Poor, and Fails to Fix the Planet*. New York: Basic Books.

Manzi, G. (2018). *Il Grande Racconto dell'Evoluzione Umana*. Bologna: Il Mulino.

McDougall, J. (1993). *Plea for a Measure of Abnormality*. London & New York: Routledge.

Mitchell, S.A. (1993). *Hope and Dread in Psychoanalysis*. New York: Basic Books.

Morin, E. (2011). *La Voie. Pour l'Avenir de l'Humanité*. Paris: Fayard.

Pellizzari, G. (2015). "Due aspetti dell'azione terapeutica: Speranza e metafora". *Rivista di Psicoanalisi*, 61(1): 157–160.

Phillips, A. (2024). *On Giving Up*. New York: Farrar, Straus and Giroux.

PopeFrancis. (2015). *Laudato Si': On Care for Our Common Home*. Vatican City: Libreria Editrice Vaticana.

PopeFrancis. (2023). *Laudate Deum (Praise God): Apostolic Exhortation*. Vatican City: Libreria Editrice Vaticana.

Portinaro, P.P. (2003). *Il Principio di Disperazione. Tre Studi su Günther Anders*. Turin: Bollati Boringhieri.

Press, J. (2019). "Psychanalyse et crise environnementale". In L. Magnenat (Ed), *La Crise Environnementale sur le Divan* (pp. 261–270). Paris: In Press.

Preta, L. (2019). *The Brutality of Things: Psychic Transformations of Reality*. K. Berri (Trans). Milan: Mimesis International.

Preta, L. (2023). "La Fantasia di auto-generazione tra diniego, illusione, speranza". *KnotGarden*, 4: 65–79.

Rich, A. (2003). *What is Found There: Notebooks on Poetry and Politics (Expanded Edition)*. New York: W.W. Norton & Co.

Ritchie, H. (2024). *Not the End of the World: How We Can Be the First Generation to Build a Sustainable Planet*. New York: Little, Brown Spark.

Rosling, H., Rosling, O., & Rosling Rönnlung, A. (2018). *Factfulness: Ten Reasons We're Wrong about the World – and Why Things Are Better Than You Think*. London: Hodder & Stoughton.

Schinaia, C. (2019). "Respect for the environment: Psychoanalytic reflections on the ecological crisis". *International Journal of Psychoanalysis*, 100(2): 272–286.

Schinaia, C. (2022). *Psychoanalysis and Ecology: The Unconscious and the Environment*. G. Lo Dico (Trans). London & New York: Routledge.

Servigne, P. & Stevens, R. (2015). *How Everything Can Collapse: A Manual for Our Times*. A. Brown (Trans). Cambridge, UK: Polity Press, 2020.

Servigne, P., Stevens, R., & Chapelle, G. (2018). *Another End of the World is Possible: Living the Collapse (and Not Merely Surviving It)*. G. Samuel (Trans). Cambridge, UK: Polity Press, 2021.

Solnit, R. (2004). *Hope In the Dark: Untold Histories, Wild Possibilities*. London: Faber & Faber, 2016.

Solnit. R. (2009). *A Paradise Built in Hell: The Extraordinary Communities That Arise in Disaster*. New York: Viking Press.

Sontag, S. (1965). "The imagination of disaster". In *Against Interpretation and Other Essays* (pp. 209–225). New York: Farrar, Straus and Giroux.

Sontag, S. (1989). *AIDS and Its Metaphors*. New York: Farrar, Straus & Giroux.

Star, Ch. (2021). *Apocalypse and Golden Age: The End of the World in Greek and Roman Thought*. Baltimore, MD: Johns Hopkins University Press.

Steiner, J. (2018). "Time and the Garden of Eden illusion". *International Journal of Psychoanalysis*, 99(6): 1274–1287.

Turner, J. & Bailey, D. (2022). "'Ecobordering': Casting immigration control as environmental protection". *Environmental Politics*, 31(1): 110–131.

Tylim, I. (2007). "Hope in a time of cholera". *American Journal of Psychoanalysis*, 67 (1): 97–102.

Tylim, I. (2019). "Hope, despair, and utopia". *Room: A Sketchbook for Analytic Action* 10(19). https://analytic-room.com/essays/hope-despair-utopia-isaac-tylim.

Ungar, V. (2015). "The toolbox of the analyst's trade: Interpretation revised". *International Journal of Psychoanalysis*, 96(3): 595–610.

Varvin, S. (2023). "Psychoanalysis and the third position: Social upheavals and atrocity". *International Journal of Psychoanalysis*, 104(3): 574–584.

Vince, G. (2023). *Nomad Century: How Climate Change Migration Will Reshape Our World*. New York: Flatiron Books.

Winnicott, D.W. (1967). "Delinquency as a sign of hope". Originally published in *Prison Service Journal*, 7, 1968. Also published in C. Winnicott, R. Shepherd, & M. Davis (Eds), *Home Is Where We Start From: Essays by a Psychoanalyst* (pp. 90–100). Harmondsworth: Penguin, 1986.

Winnicott, D.W. (1971). *Playing and Reality*. London & New York: Routledge, 2005.

Žižek, S. (1992). *Looking Awry: An Introduction to Lacan trough Popular Culture*. Cambridge, MA: MIT Press.

Chapter 2

Radical Hope

Gohar Homayounpour

This chapter attempts to elaborate the problematics of catastrophism in the face of excruciating social and political realities in Iran. Following the radical feminine uprising in Iran ongoing since September 16, 2022, the author's wish is to illustrate that the antidote to the catastrophism of our times could be witnessed in the resurrection of the erotic, a resurrection that has been at the very core of this subversive feminist revolt, of a birth of new feminine epic hero, towards an ethics of life, and its conditions.

In the very fabric of the saying "woman, life, freedom," there is clear connection to life, binding, linking, libido and sublimation [...] they are not saying we want to die for freedom, they are saying we want to live for freedom: they are taking to the streets, risking their lives towards better conditions of life. A life of dignity, pleasure, freedom and of passionate transformations, all the very derivatives of the life drive, the ethics of the erotic, indeed the enlivening antidote to catastrophism in the most devastating of conditions.

This chapter attempts to elaborate the problematics of catastrophism in the face of excruciating social and political realities in Iran. Following the radical feminine uprising in Iran ongoing since September 16, 2022, the author's wish is to illustrate via psychoanalytic clinical vignettes that the antidote to the catastrophism of our times could be witnessed in the resurrection of the social/erotic thinking subject, a resurrection that has been at the very core of this subversive feminist revolt, towards an ethics of life, and its conditions. Dreaming a dream of Radical hope, symbolized in the objectivizing function of the drive and in that of the capacity to mourn, whereas declining the invitation to the entrapment and concreteness of the melancholic discourse, and that of the superego. Attempting to gaze towards a beyond, towards the ethics of the social.

In Jonathan Lear's significant book, *Radical Hope: ethics in the face of cultural devastation*, Lear tells us that shortly before he died, Plenty Coups, the last great Chief of the Crow Nation, told his story – up to a certain point. "When the buffalo went away the hearts of my people fell to the ground," he said, "and they could not lift them up again. After this nothing

DOI: 10.4324/9781003498605-2

happened" (Lear 2006). It is precisely at this point, in the face of complete devastation, unimaginable losses and cultural collapse, of the kind we've become so brutally familiar with these days, that it becomes seductive to fall into the abyss of catastrophism. It is precisely at this significant juncture, in the face of complete vulnerability, a vulnerability indeed not foreign to the human condition, that Lear offers his vision of Radical hope. Which should unequivocally not be confused with hope.

The symptomatology of hope is a continent not too far away from the Lacanian jouissance (Lacan 1961). Essentially, if one looks at the Freudian map of the psyche, they share clear borders. For you can hope for your immortality, hope for an Oedipal triumph, hope for limitless possibilities of enjoyment and choices, hope for a re-finding of a fusional relationship with one's primary love object, hope for a life without frustration, without separation or disturbance, hope for omnipotent possibilities and completeness within yourself and the other. In short, we can say that hope is intertwined with jouissance and excess pleasure, in the territory of death, while hope-lessness has a clear link to renunciation, limited choices, freedom, pleasure, passion and life. Thus Nietzsche (1878) says: "Hope in reality is the worst of all evils because it prolongs the torments of man." But this Nietzschean foundation should not be read as a pessimistic nor a nihilistic position; on the contrary, it is a yes to life, the affirmation of an active pessimist, a passionate pessimist, for life awaits only the hopeless. Said another way: hope becomes a defense to living a passionate, connected, free, and curious life.

So, in a sense the hopeful and the catastrophizers become quite alike within the Freudian elaboration of the psyche, and both groups are not living passionately within the ethics of the social, both have become sleepless, ter-rified of their dreams/nightmares. For the catastrophizers and the hopeful refuse to mourn the inevitable loss of paradise, there is a refusal to facing one's wounds and those of the other. As a result, having fallen into the con-crete, narcissistic state of melancholia, in the absence of the capacity to symbolize, to mourn loss, and come face to face with limitations. In both categories there is an ongoing wish for the possibility of the perfect world one can envision, without disturbance and frustration, and the inescapable danger of complete withdrawal and absolute passivity in light of the inevitable disappointment of such a wish.

Radical hope on the other hand is acting with hope in the absence not just of rational justification for hope but in the absence of the conceptual build-ing blocks out of which a better future might be constituted (Lear 2006). Where we move from the moral cowardice of hope à la Nietzsche to a sense of ethics embedded within radical hope, part and parcel of an ethics of life, of the erotic, of the social. Following Lear: How should one face the possi-bility that one's world as one knows it might collapse? This is a vulnerability that affects us all – insofar as we are all humans and part of civilization, and civilizations are themselves vulnerable to collapse. How should we live and not just survive in light of or even because of such vulnerability? Can we make any sense of facing such a challenge in a way that is ethical, unpredictable, passionate and communal –all derivatives of Radical hope – or do we choose the very predictable, concrete, closed, dreamless temptation of catastrophism?

In a sense the notion of radical hope is one that is embedded within the capacity to dream, one maintained, despite imposed socio-political traumas, as in Iran for example. It is a no to the internal and external traumatizers, not out of a sense of denial nor any liberal notions of cure, but because one does not let go of the capacity to dream, to become, not despite our wounds but because of them. This is gloriously indicated in the feminine uprising in Iran, encapsulated in the clinical vignettes which will follow.

It seems that catastrophism is particularly seductive when it comes to certain geographies such as Iran: the idea of radical hope is definitely and easily foreclosed for certain lands and peoples, via both internal and external forces. As such the catastrophizer and the internal catastrophizing psychic position, become the traumatizers the moment they indicate that any possibility of radical hope is foreclosed, for example, in their consideration or lack thereof towards certain lands and people, such as Iran... The first thing that is killed in the traumatized is the capacity to dream. Today's traumatized become the catastrophizers of the future for they cannot dream things up anymore .... So, any catastrophizer is a traumatized non-dreamer, and we need to transform the catastrophe into a dream. Not out of a sense of denial or superficial optimism nor hope, but as an invitation not to foreclose the sense of dreaming, out of a vision of Radical hope, that of an ethics of life and the erotic.... we are not closing ourselves in the prison of their fixed and definite narrative (for they have, a-priori, figured it all out). For then there would be no dream, no play, no curiosity and a resignation of passions and wishes and without wishes there will be no dream work in the best sense of our Freudian metapsychology.

Radical hope is ultimately the objectalising function of the life drive, it's the wish for community, a sense of sociability, the "I" that only becomes possible via the Other. Radical hope is essentially the ethics of the social. So in a sense, the catastrophizers are lurking in the land of the psychotic, at best they are paranoiacs and at worst deathly boring, non-erotic, caught up in a regression to the non-symbolic, a place with no irony nor poetry or doubt, all the derivatives of the erotic. This feminine uprising in Iran becomes a superb example of a community of non-catastrophizers, a collective not in the name of sameness but that of difference, who dare to dream in the face of complete devastations, to dream and inspire a vision of Radical Hope that Becomes-Woman/Life/Freedom. Which in final analysis is a question of ethics, of an ethics of life and its conditions not because one denies death nor the destructive capacities of the death drive, internally and externally, but precisely because one has acknowledged it within oneself and the other.

In the very fabric of the saying "woman, life, freedom," there is clear connection to life, binding, linking, libido and sublimation ... they are not saying we want to die for freedom, they are saying we want to live for freedom: they took to the streets, risking their lives towards better conditions of life. A life of dignity, pleasure, freedom and of passionate transformations, all the very derivatives of the life drive, the ethics of the erotic, indeed the enlivening antidote to the death drive, and one of its most faithful derivatives, that of decadence in the most devastating of conditions.

Death and decadence were very much encapsulated in the collapse of the Metropol building in Abadan in the province of Khuzestan in Iran on May 23, 2022. A metropole, incidentally from the Greek metropolis for "mother city" is the homeland. Decadence comes from the Latin verb *decadere* meaning "to fall" or "to sink," which verb is also the source of decay. I would propose a more apt meaning of decadence to be to "collapse," as did the ten-storey building in Abadan. Where the building was legally allowed to have only six storeys, four additional floors had greedily been added during construction, leading to a collapse that had lethal consequences for many.

Collapse is the inevitable destiny of the subject caught within a non-discourse of decadence; social, political and cultural collapse is bound to follow. Not as the result of a fallen subject barred by the lack of being (Lacan 1961), for as psychoanalysts we know that we are all, inevitably, that. It is the subject who has collapsed into themselves. The subject that has not become social, the non-ethical Subject. For Nietzsche, decadence is more than a propensity for sexual excess or luxurious indulgence: its existence reveals a fundamental level of human disunity. 'Nothing has occupied me more profoundly than the problem of decadence' Nietzsche writes in his preface to *The Birth of Tragedy and The Case of Wagner* (1967). This unbound decadence along the lines of the Freudian death drive, beyond the pleasure principle, is bound to collapse unless there is an intervening of an ethics of life. Unless Eros speaks and invites the silent Thanatos to a duel, and this is precisely what has happened in the event of this feminine uprising in Iran.

A series of protests and civil unrest began in Iran on September 16, 2022 as a reaction to the death of Mahsa Amini, after she was arrested by the morality police for wearing an improper hijab, in violation of Iran's mandatory hijab laws.

But I will not speak on their behalf, for who should speak for whom? They have made it blatantly clear that they do not want to be spoken for, as I am sure you have heard by now…here are just a tiny representation of their own words, in their own voices, which I as a psychoanalyst on the ground in Tehran had the privilege of hearing and making an attempt at inscribing. Not in the classical psychoanalytic format, although that would be significant, but here my attempt is to highlight that the antidote to catastrophizing is indeed Radical hope, it is the erotic ethics of the social. It is precisely at this point, in the face of complete devastation, unimaginable losses and cultural collapse, that we can hear in every single one of their voices what it means to refuse the foreclosure of Radical hope, to make the ethical, communal choice of refusing to collapse or to arrest their dreams.

Although I previously wrote on this Feminine uprising, I declined to publish these clinical vignettes at the time of the ongoing protests, when everything was happening in vivo, raw and visceral. Everything was "in" action. I believe now the time has come, "afterwards" to transcribe these voices forever, for our generations to come, now as never before we need to authentically recognize that Radical hope is not to be foreclosed, as seductive as that might be at times. To be ethical subjects we must internalize Radical hope, not in the name of "reckless optimism" to borrow Arendt's term, an optimism that for me recklessly refuses to mourn our lost paradises, forgetting Proust's (1927) assertion that the only real paradises are the ones we have lost. Now more than ever in my geography and yours, where we are witnessing, wars, climate challenges, genocide, fascist political awakenings, famine, the crisis of immigration and refugees and conflictual technological advances. We need to endeavor to internalize radical hope in the name of the ethics of the social and attempt to understand from a psychoanalytic point of view what might get in our way of doing so.

So let us hear them...

A young girl that came to her session from the street said:

> "I told my friends, will it be ok to leave you for an hour to go to my psychoanalyst? They said, absolutely you will just be continuing to do what we are attempting to do here, this movement is about the birth of a subject, you will be doing this exact thing on the couch, laughing they said although not in such an effective and revolutionary way, no offence to the Freudian revolution..."

Another veiled and devout patient came to her session saying:

> "You know better than anybody that I am part of this struggle, all these years the whole discourse of my analysis has been about exactly what this uprising is elaborating, it has nothing to do with the fact that I choose to be veiled, I stand with them, for the future of my daughter and that of this land depends on it..."

Another patient said:

> "We have strategies to protect ourselves: today we asked people to join us in squares where there are no guards and oppressive forces, maybe we will not find such a square, in that case we will make our own squares in little streets and corners, we will find our unguarded squares. We also

protect each other, we are continuously thinking and we refuse to be derailed by any particular group's agenda, outside or inside of Iran, we are not fueled by superficial noise, but we acknowledge an authentic sense of solidarity, we do not have the idea that everyone should come to the street as some of us have chosen to do for that would be an identification with the discourse of the aggressor, we have understood the difference between an oppositional rebellion without a cause as opposed to a thoughtful, long and patiently awaited and created process towards a revolution, we are patient, we are not impulsive...you know it all started with the magnificent symbolic gesture of Vida Movahed but no it started even before that, and certainly after that, for that event did not happen in a vacuum but in a specific time and space that was the consequence of various other interrelated events, Vida Movahed encapsulated something, in that "still" moment of revolt."

A male patient came to his session, he the torch carrier of the obsessive curse of the impossibility of desire, and said:

"The whole time I have been here criticizing my adolescent daughter for being a lost generation of social media, the regression of images, the autism of zoomers, but alas I never thought about how maybe being born into the age of social media instead of learning about it as I did, has also been a blessing for her and her generation, they are not isolated even in a country that has been isolated in many ways from the rest of the world and yet they have been on the margin, this very position, this very inbetween position has allowed them to develop a thoughtful discourse of revolt and desire, look at them, look at what they are doing. Maybe there is a lesson to be learned for the West in regards to the crisis of adolescence, that we hear about continuously. My daughter's generation in this specific geography had no choice but to develop a political conscience, of their inherent relatedness to each other, to others, and of their own sense of alterity, they are much more aware of the outside world, my daughter the other day said I am also doing this for my Afghan sisters, maybe even the rise of the Taliban in our neighboring Afghanistan has affected and contributed to what is going on today in Iran."

Another female patient said:

"I am in the streets with my brothers and sisters not even because I am political, at least not in the clichéd sense of the word, but because I want to live. This is an uprising against death, this is a call for life, I do not want to die in the name of freedom but in order to live a pleasurable life I must join this movement."

Another male patient who continually goes to the streets to help the people who get injured while protesting, as he is a medical doctor, said:

"You know I follow two women on twitter who are the daughters of famous martyrs of the Iran/Iraq war. They are veiled and come from very political families who believed in the ideals of the 1979 revolution, well their discourse is magnificent, they not only stand with this current movement but they bring something new to it, and they are embraced by the girls who look very different from them, who had lived very different lives from them…you know what they bring reminded me of what you quoted from a famous Islamic philosopher in your book, he said he is optimistic about Iran for if we look at many countries in our region, they are at the beginning of the dream of an Islamic state, we are at the end of the nightmare…these daughters of the martyrs represent that very discourse."

Another young girl came to her session and said:

"This is not an ideological movement, I despise it when some people, usually the ones not directly involved, make it about ideology, or they make these dangerous splits and \categorizations, and namings." "They say oh that actor needs to be boycotted or that football player is persona non grata, for they do not support us exactly the way we would like. To dictate to them to show their loyalty to us, though we do not know anything about the specifics of their lives and situation. The moment we dictate to anyone how they should be we are becoming like the very people we are opposing. But I can assure you that these voices are not representative of our uprising. They have their own agendas."

"We are for all, even the guards: you know not all of them are perverse motherfuckers, although some are… the other day I went to a couple of them to rescue a girl I did not know but they had caught and I said you are tired, we are tired, just let her go please. He looked at me and said I am indeed really tired, just go home both of you, take her and take care. I despise it when they turn our sophisticated, thoughtful discourse into a discourse of good and evil, they cheapen us. This movement inherently stands for all."

A middle-aged woman comes to her session and says:

"As I was coming here there was a great deal of commotion and honking (honking has become a way of showing support), and I thought on getting closer I would see a large group of girls. Then I saw it was just one single girl about 16–17 years old standing on top of a car waving her scarf, a scarf made up of the material of flags, she was single handedly waving her newly acquired symbolic flag in the air, smiling, and what a smile it was…. I do not do much except go veilless everywhere…. I told my niece the other day that I feel guilty, and helpless. She said: Don't you go to psychoanalysis? Guilt will not help us, it is sort of useless auntie, and we are all feeling helpless at times too, that is part of it all, find a way to name it, don't get rid of it…we are naming it…"

A female patient who is a psychoanalytic candidate came to her session and says:

"On my way walking here, a woman stopped me and said, I am scared to join you all on the streets and I continuously feel shame, but I am just a really scared person in general, what can I do to help?" I said to her first of all you have every right to be scared, nothing to be ashamed of, we have space for scared people in our uprising, there is space for every-one…our uprising is about a commitment to life… if you work through your shame which has probably accompanied you well before this upris-ing, which might have to do with your feelings of being a desirous sub-ject then you have contributed to our mission, maybe that is what you need to find the courage to work through and in doing that you become part and parcel of us, of this land."

A veiled patient who has been continuously active even before this uprising against obligatory Hijab laws, comes to her session and says:

"I have a concern I feel I cannot discuss at this particular moment any-where else, you know I worry about Islamophobia, my sister who is veiled and a medical student in the states has been targeted more at her university since our uprising here. I know very well that this is not representative of the ideals of our movement, she and I are part of this uprising, at my university here, I am the one that goes on top of a stand and shouts 'Woman, Life, Freedom; and our unveiled and veiled sisters follow me…but in the discourse of Islamophobes this could be misused, and I worry about that, but I fear to discuss it on my social media at this particular time, as I don't want to be misunderstood."

A male patient who is very supportive of the current uprising, comes and says:

"My wife is against this uprising, although she is not veiled and she is not happy with the current government, her fear is that if there will be no obligatory Hijab laws, all these beautiful girls will be seen, with their perfect bodies and stunning hair and she will be even more excluded and that I will be stolen by these girls…her concerns are quite parochial, can you imagine she says yes to so much oppression just because she feels competitive with these young, beautiful women, and then she says 'Have you seen them they have no shame like my generation about being a woman, about wanting what they want, about revealing their bodies and desires, how can anyone resist them?' These women are all just irresistible."

A patient who is seven months pregnant came to her session and said:

"My husband was telling me this was a really bad timing on our part to bring a little girl into this world, the times are too turbulent. I told him I feel the exact opposite, I am delighted that my little girl will be born after such an uprising has taken place, that she will be born into a heritage of this incredible feminine uprising. You know the losses have been great, every time I think of the unbearable pain of the mothers and fathers of our brothers and sisters that we have lost I feel a pain that is extremely difficult to verbalize, every single one of their faces has become the stuff of my nightmares and dreams combined, as I am working hard at giving birth to our next generation of comrades. The other day I went for my ultrasound appointment and as I was listening to the heartbeat of my unborn daughter I thought that her heartbeat is the sound of the heartbeat of all those that we have lost in this uprising, she comes from that heritage, she will carry their heartbeat for them, their memories inscribed upon her body, not just the very unreliable historical memory but that of our bodies, that of my daughter's body, their name is like the blood in her veins that will pump her heart and with every heart beat in that of many generations to come they will be alive, mourned and celebrated. So, you see this is a wonderful time to have a baby girl after all."

A young woman patient, who is in psychoanalytic training comes to her session and says:

"I know we are continuously mourning so many losses of this uprising, but I can't help but have a celebratory mood, we certainly mourn together but we have not fallen into melancholia, we are too linked to each other, to the ones we have lost and to ourselves and to the outside world to become melancholic. If you go outside around the time that schools close, you cannot but help to want to celebrate, all these young students, veilless, hand in hand with their mothers and grandmothers, who go to pick them up also without a veil, eating ice cream and oh my, don't you think the very discourse of the transgenerational dynamic between women are changing in this land through our uprising. It is as if for the first time I am not experiencing my mother as a prohibiting, maternal superego but one that says yes to my femininity and sexuality, and through that process of course to that of her own. We feel separated, not inside each other, in an intimate togetherness that I have not felt before.

I was thinking when we encounter other women in the streets these days it is not to check her out to see if she is pretty or not, or to bestow judgments upon her style, the symbol of a competitor, we smile at each other, knowing very well that we are deeply interconnected, stronger together, and we are each grateful to the other, we are so libidinally connected, maybe we were all along, but it is as if we are now not scared to feel that connection, we can put it on full display and feel a strength within our sense of self, like never before.

I am an analyst in training, I am not saying all jealousy and envy will be eliminated, and we know very well that the birth of the subject is not smooth sailing, it is turbulent all the way, so I am not idealizing anything. All I am saying is that there is a new discourse being formed, a disturbance in the chain of the signifiers within the relationships of women and our transgenerational lineage."

A female university student, comes for her session after a few particularly tumultuous days at her university and says:

"You know if things don't start at universities where should they start from? I can assure you that we will go on, isn't it funny that we are the children of this revolution, most of us that are participating in this uprising are born after the revolution. You know my older sister who is in her 40s said to me the other day, your generation is different from our generation. I am not sure what happened, but I cannot imagine any of us taking our hijab off or even telling off (we have to keep telling-off) the school or university authorities. My sister went on to tell me a memory that once, when she was in high school. told our mom that the head-mistress told her she will go to hell for I had told her it is not my desire to listen to her nonsense, I wanted to have boy classmates, mom said just be quite at school, it can be dangerous for our family. And then you came and for years now you have been revolting and no one can stand in your way. I envy you, I admire you…when I see the videos and photos of all that you are all doing, it is as if I get to live my dreams through it and now when my 13-year-old daughter came home and said I am not wearing my hijab to school anymore I was tempted to repeat to her what mom told me, to just wear it and be quiet. But then I remembered you, all of you and thought, through your courage I will find the courage to tell her otherwise, so I said go for it, let's do it together, we can join your aunt and her friends."

Another female patient comes and says:

"Something very interesting is happening that goes beyond the usual discourse of these days, my ten year-old daughter came home the other day and said our gym teacher all year long has been saying, football is not for girls, you have to participate in a softer sport and none of us liked it, but we did not say anything and now we made an uprising of our own and said to her together why do you say that? We want to play football too, we will show you how good we can be, just give us a chance, which she did. And the other day I was doing her science homework with her and there was a picture of prehistoric men, and she said 'There was no prehistoric woman then? Why is there never a picture of her?'

In general, I have noticed since this uprising in the country, she and her friends are continuously discussing girls this, boys that, or that's not fair. I know that at some level it could be her age, and many other layers of her becoming. But I really think that the outside uprising has penetrated into their discourse, all these little girls are asking new questions, thinking about things in a new way, isn't this marvelous. This uprising has given generations of girls of this land the chance to think in a new way, isn't this marvelous?"

In these clinical vignettes, it becomes blatantly clear that operating within the nucleus of this uprising is an essential function of the life drive, the objectalising function of the life drive à la Andre Green (1999), where they are capable of transforming structures into objects, even when the object is no longer directly involved. In the very fabric of the saying "woman, life, freedom," there is clear connection to life, binding, linking, creativity, and sublimation. The very opposite of Bion's "Attacks on linking":

"The destructive attacks which the patient makes on anything which is felt to have the function of linking one object with another."

(Bion 1959)

So not attacks on the object itself but on linking.

They are marching towards the ethics of the social, with an inherent sense of attentiveness and basic trust towards the next generations to come, a clear transgenerational lineage, relentlessly gifting their gaze upon Radical hope, which a priori requires an authentic internalized awareness of their vulnerability, of interconnectedness to each other and to all others both far and near, in all geographies and of a desire for cohabitation, for intimacy, and for the erotic.

For to survive a life is drastically different than living a life, and living a life is only possible via a becoming of the subject through an awareness of the other. We know very well that this route has its own set of hells and problematics but it is none the less the only possible route towards becoming a thinking subject. For our lives begin to end, even if we survive, the second the life and conditions of life of another are not treated with dignity and grievability. For as Judith Butler (2004) reminds us: "To cohabit the earth is prior to any possible community or nation or neighborhood. We might sometimes choose where to live, and who to live by or with, but we cannot choose with whom to cohabit the earth."

This is also where psychoanalytic ethics comes from, it can only come from a social/communal, thinking subject, in other words from a sense of separateness, of a distance between I and the other. It is only through un-belonging that we are granted the possibility of becoming, of subjectivity, the "I" that can only become via the social, taking the bite of the apple and being thrown out of paradise. Is it not true that Adam and Eve became neurotics as opposed to the psychosis that would inevitably afflict them if they stayed in paradise forever? Their desire for knowledge, their sin is ulti-mately the wish to know themselves and the Other, and it is what takes them out of the dangers of primary narcissism, and throws them outside of para-dise, into the outside world, where they will encounter the Other and them-selves. It is the moment they become social beings... In our psychoanalytic terms it is what does not happen that leads to psychosis as their sense of the social is blocked, they stay merged with the maternal object, one and the same, in Lacanian terms with a foreclosure of the paternal function...stuck in the dyadic imaginary paradise of the primary object. Indeed, our need for belonging will inevitably block our way to the Other, to the outside world (Homayounpour 2024). It will interfere with our way towards our own "becoming," towards what I have termed the erotic ethics of the social.

Hence ethics comes from a seduction that was not actualized, from long-ing, all the various possibilities when need and object are not the same anymore.

Ethics comes from seeing the other as other, ethics comes from love.

Love comes from mourning the loss of the love object.

It is not about self-sacrifice, pure devotion, or kindness for the sake of kindness, it is about becoming a loving/social subject.

Hence, we can continue to elaborate that the catastrophizer's road to the outside world has been blocked, they are somehow withdrawn melancholics, who have not mourned the loss of the other, convinced that how things are and feel at the moment, will forever remain as such. This is the essence of trauma which is encapsulated in the belief that the situation we are in, or how we are presently feeling, is going to last forever, that it will always remain as such, that it will never end.

> "Would it really be a trauma of birth if we could whisper in the new-born's ear that this will only last a few minutes? Or in the midst of depression after a devastating loss, if only we believed that things will change, that we would one day laugh again, it would not be so traumatic. Trauma is defined by its timelessness."
>
> (Homayounpour 2023a)

Can we say that catastrophizing, just like trauma is defined by its timelessness?

I am reminded of Julia Kristeva's beautiful assertion in her book Black Sun, in which, simply put, she asserts that depression starts early, and it is a pre-Oedipal pathology coming from the fact that one does not know how to lose the maternal object. "If I did not agree to lose mother, I could neither imagine nor name her." In short, according to Kristeva, one's melancholia comes from not knowing how to lose.

Kristeva makes a clear distinction between Dostoyevsky, "the writing of Suffering and forgiveness," an example par excellence of the successful sublimation of mourning, and the "discourse of dulled pain" expressed by Marguerite Duras.

Dostoyevsky is able to verbalize the effect of suffering in a new style. For Kristeva, Dostoyevsky's writing achieves the transformation of melancholia that best encapsulates how the "unconscious might inscribe itself in a new narrative that will not be the eternal return of the death drive." Unlike Dostoyevsky, Duras's "discourse of dulled pain" captures the "malady of death" in "an aesthetics of awkwardness" and "a non-cathartic literature," a style which is certainly distant from Radical hope. As Kristeva writes, "There is no purification in store for us at the conclusion of those novels written on the brink of illness, no improvement, no promise of a beyond, not even the enchanting beauty of style or irony that might provide a bonus of pleasure in addition to the revealed evil." For Kristeva, Duras's writing, unlike Dostoyevsky's, demonstrates her inability to sublimate her "passion for death." Kristeva (1989) goes on: "Duras does not orchestrate [the malady of suffering] in the fashion of Mallarmé who sought for the music in words, nor in the manner of Beckett who refines a syntax that marks time or moves ahead by fits and starts, warding off the narrative's flight forward."

I think we can easily just substitute catastrophizers for melancholics in Kristeva's discussion.

Within Freudian metapsychology we need to be fundamentally disturbed via absence in order to develop the capacity to hallucinate a breast/bottle and thus to get closer to the reality principle through which we can exit our primary narcissism and enter Green's objectivizing function of the drive.

Here we should follow Freud (1917) when, in *Mourning and Melancholia*, he elaborates on pathological melancholia as opposed to mourning. "Mourning is regularly the reaction to the loss of a loved person, or to the loss of some abstraction which has taken the place of one, such as one's country, liberty, an ideal, and so on." The same factors, Freud tells us, produce melancholia in some people.

He continues to say that in both cases, mourning and melancholia, there is a cessation of interest in the outside world, inhibitions on all activity, and a loss of the capacity to love. There is, however, one fascinating exception in melancholia: we observe a significant lowering of self-regard. This dialectic of crime and punishment is absent in mourning.

Other fundamental differences are touchingly elaborated by Freud, such as that the melancholic only knows whom he has lost but not what he has lost. This would suggest that melancholia is in some way related to an object-loss which is withdrawn from consciousness, in opposition to mourning, in which there is nothing about the loss that is unconscious. "In mourning it is the world which has become poor and empty; in melancholia it is the ego itself" (Freud 1917).

The analogy with mourning led us to conclude that the melancholic had suffered a loss in regard to an object; what he tells us points to a loss in regard to his ego. So, the melancholic's ego has split, and one part has revolted against the other. We see how one part of the ego sets itself against the other, judges it critically, and, as it were, takes it as its object. This critical agency within the ego becomes what Freud later called the superego. But upon a closer look at each clinical picture, we perceive that the self-reproaches are reproaches against a loved object, which have been shifted away from it on to the patient's own ego. Freud (1917) goes on to explain:

> "Thus the shadow of the object fell upon the ego, and the latter could henceforth be judged by a special agency, as though it were an object, the forsaken object. In this way an object- loss was transformed into an ego-loss and the conflict between the ego and the loved person into a cleavage between the critical activity of the ego and the ego as altered by identification."

This critical agency which in time will be named the superego, the memorial site of our early identifications, is where the poisonous seed of the wish for ideals and a morality that reeks of blood is cultivated. For the superego is not exactly the voice of an ethical authority, but simply that of a moral authority, and I believe this is an important and at times misunderstood notion. The superego for Freud is an unrelenting command, one that we can never satisfy: the harder you try the guiltier you become, in a deeply Kafkaesque sense. The superego is an insatiable hungry master operating within the sado-masochistic realm: the more you feed it the hungrier it becomes. The superego takes the subject's own drives and turns them against themselves. So in a way, we can say that the superego becomes a perverse agent within the ego (Homayounpour 2023b).

Hence the superego gives way to a violence in the subject, a regressive, narcissistic violence from object libido to ego libido. A violence that stokes us mercilessly, becoming an all-knowing master that we fear and must obey in order not to get to know ourselves and the sexuality and violence a part of us recognizes within.

Yet we know that the superego is a necessary part of developing a mind, as are early identifications and even the early wishes for belonging, and that we don't have a shot at becoming subjects without those, so we are in a pickle;

the psychoanalytic pun is intentional here. But, as neurotic subjects, WE ARE IN A PICKLE (Homayounpour 2023b).

The superego claims to force the ego to act morally, but not realistically, and this lack of connection to the external, outside world (to our sense of the social in a paradoxical way) is part and parcel of how the story of the superego, which started as a necessity in the history of our becomings, gets perverted. The superego is indeed quite antisocial, hence quite close to the id. Within Freudian metapsychology, the discourse of melancholia is intimately connected to that of the superego, and of a regression to primary narcissism, paving the way for the catastrophizers, the idealist and omnipotent souls of the future. Where Radical hope is refused, even foreclosed, it is a no to the ethics of the social, a collapsed subject that refuses to envision a beyond, to dream and be intimately linked to the authentic, curious, passionate encounter with different parts of oneself and that of the other. With an acknowledgment of the death drive and all its unbinding derivatives palpably roaring in every single one of us, till death do us part.

*

Radical hope is intrinsically connected to the communal, to the ethics of the social, consequently without it we will collapse as our death drive is driven to makes us do so. We are certainly mortal beings, and the turbulent birth of the subject is inevitably a fallen subject, but are we doomed to collapse? Greed and decadence are in opposition to the ethics of the social and radical hope. Can we live a life that is ethical, passionate, pleasurable and not just solely a condition of survival – that is, of the imaginary illusions and false promises of hope – but within communality, part and parcel of the thinking subject, leading to a social sense of responsibility and freedom? Which a priori requires that one develops the capacity to mourn, Catastrophizing is a refusal to mourn absence, and it's a narcissistic wish to not give up our omnipotence, nor that of the other; it's a refusal to face absence, disturbance, and limitations; it is a no to the confines of the reality principle, with all its limitations and delays. Morality certainly does not work for it strengthens the punitive superego, so intimately connected to id as explained previously.

The antidote being what is traditionally called doing superego work, is ultimately really id work. In a sense, the famous psychoanalytic clinical notion of doing superego work is really to get to know the wishes of one's id, its darkest, most hidden, unacceptable and strange longings, marching towards the becoming of the social/thinking/erotic ethical subject.

For Hanna Arendt reminds us that "a life without thinking is quite possible; it then fails to develop its own essence – it is not merely meaningless; it is not fully alive." A thinking/desirous subject is born via a linking to the other; not out of a sense of belonging which in final analysis isolates every intimate contact but towards an un-belonging. This un-belonging is the necessary step towards the becoming of the ethical subject that opens itself up to the outside world. Not confined to the imprisonments of clans, groups, families, countries, religious groups, and ideologies. For any attempt at belonging will exclude an "other," it might give us the illusion of safety, but it will inevitably end up in Dante's inferno with no Virgil in sight, as we are currently observing in various contexts in the world. Can we wake up from our current nightmare (nightmares are nothing but failed dreams) and find our way to the land of dreams? This will require radical hope, which a priori necessitates giving language back its metaphoricity, to have an internal trust that we are all interconnected, this sense of ethics is not a morality that often reeks of blood. But to come face to face with all that is unacceptable/refutable within, our darkest, oddest secrets, of our sexuality and of the death drive inevitably pulsing within all of us, on the edge of language. To move beyond good and evil, not out of any liberal sense of empathy or cure, nor the wish to heal our wounds or those of the others. But out of a sense of our interconnectedness, recognizing that this ethics of the social must underlie any liberating praxis, and to be reminded of our common human heritage, of our common fragility. This communal reminder and joint injury to our narcissism becomes a welcome messenger, a way of re-finding, in final analysis, that we are all linked, far and near, in sickness and in health.

Radical hope is a social/ethical act of defiance towards ideologies, not a rebellion without a cause but a defiance that will be revolutionary, an event that will disturb various levels of significations, internal/external, private and public, personal, political, social, and the very discourse of subjectivity. Not in the name of sameness that does nothing but exclude, but for difference, with a genuine comprehension that our emancipation is inherently intertwined.

The words of the subjects of this Iranian feminine uprising are a marvelous example of such ethics of the social, of their gaze upon their transgenerational lineage, or to use Hans Loewald's (1960) words, through the "blood of recognition." In the name of the ethics of the erotic; of life and nothing else. Heard through every single one of their voices, through their singular/collective voice. Where we truly are willing to get into the risky business of encountering the Other, with all its inevitable hells and pleasures. Can we give primacy to the ethics of the social, which is of cohabitations, of the interconnectedness of all things, or will we choose the concreteness of nightmares? Forever doomed to isolation, beyond the pleasure principle; and to never truly get a chance to cohabit different parts of ourselves, nor encounter the other and the universe in a meaningful and pleasurable way. Not in the confines of our parochial belongings but with an authentic curiosity towards the other and our own unconscious, where we move away from the imaginary comforts of a wished-for sameness towards the glorious disturbances of difference, no ifs, no buts.

For Arendt (1960) is right when she says freedom is identical with the capacity to begin. Let us do the unexpected, to begin again and again, "If not here, then there. If not now, then soon. Elsewhere as well as here" (Sontag 2007).

References

Arendt, H. (1960). "Freedom and politics: A lecture". *Chicago Review*, 14(1): 28–46.
Bion, W.R. (1959). "Attacks on linking". *International Journal of Psychoanalysis*, 40: 308–315.
Butler, J. (2004). *Precarious Life: The Powers of Mourning and Violence*. London: Verso.
Freud, S. (1917). Mourning and Melancholia. *SE* 14.
Green, A. (1999). *The Work of the Negative*. A. Weller (Trans). London: Free Association Books.
Homayounpour, G. (2023a). *Persian Blues, Psychoanalysis and Mourning*. London & New York: Routledge.
Homayounpour, G. (2023b). "The Ego and the Id and… the superego". In F. Busch (Ed), *The Ego and the Id: 100 Years Later*. (pp. 25–31). London & New York: Routledge.
Homayounpour, G. (2024). "Dear Eyal". *Psychoanalytic Dialogues*, 34(3): 317–327.
Kristeva, J. (1989). *Black Sun, Depression and Melancholia*. L. Roudiez (Trans). New York: Columbia University Press.
Lacan, J. (1961). "The directions of the treatment and the Principles of its power". In A. Sheridan (Trans), *Écrits: A Selection*. New York: W.W. Norton & Co., 2001.
Lacan, J. (1966). "The subversion of the subject and the dialectic of desire in the Freudian Unconscious". In B. Fink (Trans), *Écrits – The First Complete Edition in English* (pp. 672–702). New York: W. W. Norton & Company, 2007.
Lear, J. (2006). *Radical Hope: Ethics in the Face of Cultural Devastation*. Cambridge, MA: Harvard University Press.
Loewald, H. (1960). "On the therapeutic action of psycho-analysis". *International Journal of Psychoanalysis*, 41: 16–33.
Nietzsche, F. (1878). *Human, All Too Human: A Book for Free Spirits* A. Harvey (Trans). Chicago, IL: Charles H. Kerr & Company, 1908.
Nietzsche, F. (1888). *The Birth of Tragedy and The Case of Wagner*. W. Kaufmann (Trans). New York: Random House, 1967.
Proust, M. (1927). *In Search of Lost Time, Volume VI: Time Regained*. T. Kilmartin & A. Mayor (Trans); D.J. Enright (Rev). New York: Modern Library, 1999.
Sontag, S. (2007). *At the Same Time: Essays and Speeches*. New York: Farrar, Straus and Giroux.

Chapter 3

A Changing World

Psychoanalysis between Catastrophe and Hope

Alfredo Lombardozzi

Psychoanalytical Anthropology of the Catastrophe

I would like to begin with the phrase: "Imagining the future." I would say that it exposes us to a work of some commitment at this time when we paradoxically find ourselves experiencing a living in the present that also has a "halo" of the future. I mean that, especially for those of us who lived at the turn of the century, which coincided with the transition between two millennia, the powerful acceleration we are witnessing on so many areas of existence (economics, technology, modes of human relations in the narrow sense or more generally geo-political, climate change, pandemics predicted but not expected, etc.) has us breathing in the air accelerated particles of the future floating around us. I have the feeling that rather than Imagining in the best-known sense we find ourselves having to reorganize streams of imagination.

We then tend to catch, albeit in some ways unprepared, the anticipatory signs of catastrophe as the future already propels us forward toward outcomes that can both exhilarate and frighten us. How, however, can we not be overpowered by the sense of catastrophe and keep hope alive? How can we "think" as the day's subtitle says, counteracting the tendency to denial, avoiding flight and retreat from a terror that is impossible to contact, from the sense of catastrophe and, above all, from the impossibility of orienting ourselves with respect to all that is happening?

I would first like to propose beginning with two "classic" psychoanalytic summits that allow us to offer some form of framework. We are talking about Wilfred Bion and Erik Erikson two very different but both very significant authors in psychoanalytic thought. I am referring to "Catastrophic Change" in Bion and the idea of "Hope" in Erikson. In different ways both concepts lend themselves to inspiration both in the analytic situation and in the interweaving of psychic reality with the external world.

I would like to try to integrate the two different models. Bion for his part describes the psychoanalytic process in terms of transformations by taking as an example, in addition to the change of a landscape in invariance, the transition from pre to the post-catastrophic. This transition in the analytic

DOI: 10.4324/9781003498605-3

relationship exposes one to the risk of catastrophe unless, in the accompaniment of an "adequate" analyst, the experience of "catastrophic change," that is, of experiencing a "controlled" catastrophe, is allowed. A transformation then takes place, involving the opening-up of an area of interest/action between the conscious and the unconscious, which also involves external reality, and enters full-circle in the process that Bion identifies in the T-transformation, whether alpha or beta, related to the apex of the "Public," which, in my view, represents "Social Culture" (Bion 1965).

From another point of view, Erik Erikson proposes an approach that unveils a more intrinsic and explicit relationship between external reality, psychic reality, drive processes and developmental contexts, such that he describes mental states not only in terms of psychosexual figures but also in the form of "feelings," which, precisely as such, are susceptible to recognition on the social level. For example, he places the coexistence of despair and disgust in the elderly as a precondition for "Wisdom," or of trust and distrust in childhood as a precondition for "Hope" (Erikson 1978).

The "Public" factor in Bion is declined by Erikson in Erikson's recourse to the value of the historicity of experience, which he well describes in a passage in his essay, where he takes his cue from Ingmar Bergman's film "The Place of Strawberries," to delineate the process of human life cycles as he articulated them. The "characters" in the film, like those in life, live in a culturally and historically determined sense of shared "earthiness" (Erikson 1982).

I would define Hope, by bringing the two visions, Bion's and Erikson's, closer to that approach inspired by Kohuttian thought, which identifies a vital factor in the drive for life and an existence of full dignity (Chiavegatti and Di Luzio 2023).

This state of mind finds its expression precisely on the "crest" of the transformative process that occurs in the transition between the pre- and post-catastrophic. The sense of Hope takes shape and strength precisely in the course of this process, in the oscillation between distrust and trust, through a psychically and culturally recognized accompaniment.

Both discourses and thought patterns refer in some ways to the well-known formulations of ethnologist and historian of religions Ernesto De Martino concerning his posthumous notes on *The End of the World* (1997, 2016) and the distinction between what he calls "Cultural" versus "Psycho-pathological Apocalypses" (1964, 1997). Already in the course of his historical-critical and ethnographic research De Martino had identified the "Crisis of Presence" as an individual and collective "state of mind," which is almost an obligatory passage. The crisis of presence is considered a crucial experience when in the experience of existing in a world and in a culture that represents it, a gap opens up that exposes the individual and the group to a sense of crisis with respect to the possibility of being in that world in a way that we psychoanalysts would call sufficiently integrated. Consequently, crisis occurs when the possibility of perceiving oneself as belonging to a culturally and historically connoted world is lost. Ritual represents the possibility of denouncing this condition of de-storification and, at the same time, resolving in the sharing of it (De Martino 1948).

The crisis of presence is subsequently declined by De Martino in the experience of the "End of the World" which leads back to the different apocalyptic visions. The apocalypses of the West tend predominantly to take the form of "Psychopathological Apocalypses," which do not involve a redemption (an eschaton). The individual implodes in his or her private suffering by not finding a foothold in some 'public' mythical ritual context, which, conversely, would be the solid ground for being able to access the "Cultural Apocalypse." The latter produces a movement that places the individual in close contact with a community that recovers the sense of history, Culture and allows the domestication of objects and affects which represents the other side of the experience of psychopathological estrangement.

It seems clear to me that these reflections and the concepts derived from them, explicated by the three great classics; of psychoanalysis Bion, anthropology De Martino and the intersection between the two, in his own way, Erikson, are in strong correspondence with each other. They are united by the courage to look into the eye of the storm of catastrophe without falling into the pit of catastrophism but proposing ways out that always keep Hope alive. At the same time, they share a sense of actuality and a constant attention to the dimension of the future. They also introduce and anticipate processes and developments that disorient us today and justify recourse to further "grids" of knowledge.

Contemporary Form of Catastrophe

There are many aspects of the reality in which we are immersed that expose us to a sense of catastrophe today. Certainly, the latest pandemic and war-related developments count, whether it be the Russia-Ukraine war or the more recent deflagration of the endemic conflict between Israelis and Palestinians. Moreover, we experience the unsettling feeling that we cannot predict developments, even if they are glimpsed, on the geopolitical level globally. Not to mention the climate crisis, which is "under everyone's eyes"; even those in denial and unwilling to see it. Finally, I would point to the development and acceleration of the power of media communications, which significantly alter the nature of human relations.

There would seem to be all the conditions for falling into the temptation of "catastrophism," that is, for submitting to that view of existence that takes for granted that we are running toward the ravine like rats following the tune of the Pied Piper of Hamelin.

We therefore have two types of reactions to the uncertainty and precariousness we are experiencing. On the one hand, to retreat into an "apocalyptic" vision, in a sense opposing reality, and counteracting it in a "nostalgic" attempt to make it fit into one's own, and perhaps in some respects obsolete epistemological and/or ideological paradigms in a form of denial, and on the other hand "to close one's eyes," or even shifting one's gaze away from the evidence of the dramatic nature of some very worrying developments in the contemporary world, in a form of defense closer to denial.

It seems to me that it is important to find a measure with which to be able to better focus on the context of what we are talking about. There is no shortage of strongly critical views of developments in current society that put "the contemporary" at the center of their attack, highlighting catastrophic drifts. On the other hand, there is a different attitude which highlights the presence of complex factors and pays attention to the convergence of different elements: psychological, economic, power and deviant management of the relationship with nature or new communication media.

Among the contributions that go in the last direction I would suggest considering some recent reflections on the anthropological, sociological and philosophical level that, in my opinion, differ from the more well-known denunciatory positions of Korean philosopher Byung-Chul Han (2013, 2019), which, although in several respects shareable, present some simplifications in supporting a vision oriented to attributing predominantly negative valences to the relational models of contemporaneity.

The reflections of some scholars who attempt to grasp and explore the current moment seem to me to be of considerable interest. For example, the sociologist Richard Sennett, who is very critical of contemporary capitalist development but not reductionist in economic analysis, had already in the past denounced the limitations of the "flexible" man, or the risk of the loss of the public sense of roles, and the recourse to an 'excess of intimacy' as a dimension that exposes one to excessively narcissistic experiences and the undermining of the possibility of sharing a sense of community (Sennett 1977).

Sennett described, with a critical spirit but with an eye always ready to grasp ways out, these processes as part of the design recovery of a constructive sense of valuing collective life, or of ritual processes, which in "repetition" allow gradual transformations in the everyday. The construction of cities and their public spaces, the denunciation of some of the significant limits of networked communication that privilege, through well-defined management choices by large platforms, polemics rather than dialogic conversation; or the recovery a "craft" way of entering into relationship with the object of one's work, which can represent from our point of view the depiction of an internal object that has its own value and constancy (Sennett 2012) are all elements of a critique of contemporary society not obscured by catastrophist sentiments.

Something similar can be said about the recent formulations of anthropologist Arjun Appadurai, who, after proposing in the 1990s an in-depth analysis of media landscapes and scenarios connected to the identity processes and social imagination that characterize them, has more recently placed the theme of "failure" at the center of his reflection in a book written with Nina Alexander (Alexander and Appadurai 2020).

The model of "failure" implies a series of convergences in the social construction of the idea of the media object and the instruments, which is the medium for networked communication, and is conceived as part of a functional structure assimilable to the system of neo-liberal finance; conforming to the principle that failure is the condition for innovation in media technology and motive for profit in the field of finance. It is, moreover, a system that is based on failure as a drive for development and consumption and implies a policy based on oblivion, or rather denial of the fallibility of the system itself that is instead perceived as highly functional. On the contrary, it fosters through refined forms of manipulation, the individual's self-attribution of responsibility for the system's malfunction and sense of failure in the relationship with media instruments.

The sense of loneliness and withdrawal into a kind of individual and private province, devoted to failure and denial, with strong repercussions on the level of life and psychic dynamics, constitute phenomena largely influenced by the global context and the power cultures it produces.

Writing Catastrophe against Catastrophism

We can reflect on the fact that in contemporary times the distinction that Heinz Kohut (1977) proposed between the culpable man, that is, the man of drive conflict, and the tragic man, suffering from narcissistic vulnerability and relational deficit, converge in an unprecedented and complex coexistence. Fragility and extreme sensitivity with respect to narcissistic wounds become intrinsic aspects in the relationship with highly conflicting elements on the intrapsychic level and in social relations.

Thus, we find ourselves in the midst of processes that foster forms of forgetting and denial, both on the individual and social level, as a defensive reaction to problems that cannot be dealt with, or are perceived such as, which are related to the rise of a culture of "susceptibility" (Flasspöhler 2021).

This leads to favoring a heightened ideological polarization of opinions rather than our encounter or dialogue, or to rigid positions that re-propose conflicts, as is the case with climate change, between those who deny the existence of the problem and the responsibilities of human intervention and those who instead propose radical solutions that do not take into account the limits within which one can still act by setting reasonable goals.

In this regard, writer Amitav Gosh, in his fine book on the relationship between literature and climate change, *The Great Derangement* (2016), denounces the lack of writing or of an epic of nature as a shared human environment. Writing can, in these terms, be seen as an attempt to counter the tendency toward denial of the serious climate crisis which is as much present to our sensibilities and in our now daily experience as it is invisible and relegated to dissociated areas of existence and the collective mind. A collection of short stories, edited by John Freeman, of writers from around the world (Asia, Africa, Europe, and America), *Tales of Two Planets* (2020) testifies, in synergy with the need Gosh denounces to counter denial, to the impact of climate change and environmental crisis on different forms of existence: cities, rivers, deserts, and oceans change their balances; territories, which in individual memories and experiences were associated with familiar atmospheres and climates, become foreign and unsettling; heat waves, hurricanes, plastic islands in the sea and rivers confront us with a polluted landscape that becomes an environment which in the not too distant future is no longer sustainable for human life itself. The narrative of a multi-voiced volume, like a complex and multifaceted mosaic, becomes not only a denunciation but also an attempt to give cultural form to a catastrophe that is no longer merely announced.

The catastrophe of a man's private existence, which drives him to retreats into himself and his own psychosis, is narrated very evocatively in an epic, dramatic and ultimately liberating tale by Stefano Valenti entitled, *Cronache della Sesta Estinzione* (*Chronicles of the Sixth Extinction*) (2023). The protagonist in the course of his life marked by traumatic events and relationships enters a psychopathological vortex, a daily apocalypse of body and mind, which involves the image of the world in its entirety and, at the same time, resonates in a kind of jubilation of fragmentation. Existence, as private and personal experience enters into an undifferentiated synergy with the world: the city, the objects, the camper that becomes a home, the sea, the river, the mountains, the tree. Everything, every element becomes a motif to trigger unpredictable processes and disturbing metamorphoses. Every internal and external movement exposes him to a dizzying fall and, each time, he encounters a void in this fall, a black hole, the darkness of those he calls "the sunken."

In this catastrophic vision there is in the background the figure of Robinson Crusoe on "his" island. An island of "light," the landing place of a castaway who enacts all his "Culture" to govern fear and loneliness; he is an antidote to catastrophe, a survivor.

Catastrophe closely relates the "monster" in him (the protagonist), as much incorporated by reality as refused and rejected (Pallier 1990; Tagliacozzo 1990), to the extinction of the world itself and of the human species, the sixth; the one that has yet to happen and is already in part happening.

A dog appears at the end of the story in a dark but "calm" environment, an if only slight respite, a claustrophilic environment that opens to a faint glimmer of hope. The dedication at the back of the book reads: "To Myself. To the reborn."

Literature provides us with two significant examples that, when read with a psychoanalytic lens, point us to the possibility of experiencing catastrophic change, that is, in the case of writing, the attempt to emerge from catastrophe through storytelling, or the possibility of strong sharing in the group dimension. If we add an anthropological lens, following De Martino, the "Psychopathological Apocalypse" can evolve into a "Cultural Apocalypse," at least in the psycho-cultural construction of a communal "domestic" horizon. In the opening of a dialogue in the passage between the different figures of catastrophe, a space is created, the ground on which the figure of Hope takes shape, a feeling that allows the maintenance of what Appadurai calls the "capacity to aspire, cultivated in the daily construction of the future. It is a cultural capacity whose form is almost "universal" and, at the same time, "local" (Appadurai 2013).

References

Alexander, N. & Appadurai, A. (2019). *Failure.* Cambridge, UK: Polity Press.

Appadurai, A. (2013). *The Future as Cultural Fact: Essays on the Global Condition.* London: Verso.

Bion, W.R. (1965). *Transformations: Change from Learning to Growth.* London: Heinemann, 2013.

Byung-Chul, H. (2013), *In the Swarm: Digital Prospects.* E. Butler (Trans). Cambridge, MA: MIT Press, 2017.

Byung-Chul, H. (2019). *The Disappearance of Rituals: A Topology of the Present.* D. Steuer (Trans). Cambridge, UK: Polity Press, 2020.

Chiavegatti, M.G. & Di Luzio, G. (2023). *Il diritto di Esistere. Scritti sulla Ricerca di Lydia Pallier.* Rome: Avio.

De Martino, E. (1948). *Il Mondo Magico. Prolegomeni a una Storia del Magismo.* Turin: Bollati Boringhieri, 2007.

De Martino, E. (1964) "Apocalissi psicopatologiche e apocalissi culturali". *Nuovi Argomenti,* 69–71.

De Martino, E. (1977). *The End of the World: Cultural Apocalypse and Transcendence.* D. L. Zinn (Trans). Chicago, IL: The University of Chicago Press, 2023.

Erikson, E.H. (1978). "Reflections on Dr. Borg's life cycle". In *Adulthood: Essays* (pp. 1–31). New York: W.W. Norton & Co.

Erikson, E.H. (1982). *The Life Cycle Completed: Extended Version with a New Chapter by Joan M. Erikson.* New York: W.W. Norton & Co, 1997.

Flasspöhler, S. (2021). *Sensibili. La Suscettibilità Moderna e i Limiti dell'Accettabile.* T. Isabella (It. Trans). Milan: Nottetempo, 2023.

Freeman, J. (Ed.) (2020). *Tales Of Two Planets: Stories of Climate Change and Inequality in a Divided World.* London: Penguin Putnam.

Ghosh, A. (2016). *The Great Derangement: Climate Change and the Unthinkable.* London: Penguin Books.

Kohut, H. (1977). *The Restoration of the Self.* New York: International Universities Press.

Pallier, L. (1990). "Il bambino 'mostruoso' come minaccia all'integrità del sé". In C. Neri, L. Pallier, G. Petacchi, G.C. Soavi, & R. Tagliacozzo, *Fusionalità. Scritti di Psicoanalisi Clinica* (pp. 147–153). Rome: Borla.

Sennett, R. (1977). *The Fall of Public Man.* New York: W.W. Norton & Co.

Sennett, R. (2012). *Together: The Rituals, Pleasures and Politics of Co-Operation.* London: Allen Lane.

Tagliacozzo, R. (1990), "Il bambino rifiutato: Falso sé, mantenimento e rottura; angoscia del vero sé". In C. Neri, L. Pallier, G. Petacchi, G.C. Soavi, & R. Tagliacozzo, *Fusionalità. Scritti di Psicoanalisi Clinica* (pp. 131–142). Rome: Borla.

Valenti, S. (2023). *Cronache della Sesta Estinzione.* Milan: Il Saggiatore.

Chapter 4

Catastrophic Fossil Culture and Other Desires of Relatedness

Mauro Van Aken

Introduction. Emergencies, Including Generative Ones

"The climate issue is driving us crazy!" (2017, 77), Bruno Latour repeatedly exclaims, starting from the emotional dimensions of "paralysis, anguish, guilt, helplessness" that make up the frenetic paralysis in the Anthropocene age: everything seems to be speeding up in the absence of any maps and visions of shared imagination of the future. The climate crisis stands out as a cultural "unthinkable": it is a dimension of an accelerating environmental change that is shattering ideas of mastering nature and the cosmology of modernity. We are failing to socially process an environment that no longer marks the seasonal transitions, nor those interdependencies with subjects and forces of the living in the midst of accelerated change and "emergency": emergency not only as an ecological crisis, but also as that which "emerges," a revelation that resurfaces and manifests limits, relationships and nonhuman agencies on which we depend and to which we belong, including emotionally.

The future, which was once a progressive line, now appears as a blockade looming over the new generations. And we remain dumbstruck because we have lost the cultural frameworks needed to collectively work out an ecological transition that is first and foremost a cultural transition, of the cultural imagination of the meanings of human and metaphors of the world, more than merely the techniques which are already available. Decarbonizing the economy can only start by a value and cultural conversion: by decarbonizing the cultural imagination of the world, of human being and the living as well as its relatedness.

A burdensome public silence therefore hangs over the climate crisis in its cultural aspects, amidst the deafening bass drum of green neoliberalism, climate policies as techno-economic, managerial processes and expert's knowledge with their attached technophanies; while forms of social activism "stage" scenes of "there is no time" emergencies and trauma. Apocalyptic, collapsing or dystopian narratives have long been prevalent in a capitalist realism that seems to have saturated every space of the possible: whether infinite growth and economization of life or its apocalyptic damnation. It is the ancient historical dichotomy of our cosmology in crisis (Van Aken 2020).

DOI: 10.4324/9781003498605-4

We have lost or transcended a shared social alphabet since we lack the frameworks needed to symbolize these changes and their nonhuman actors with the limitations and relationships which they impose. As Amitav Ghosh writes, "a broader imaginative and cultural failure that lies at the heart of the climate crisis" (2016, 8), precisely because, with the illusory power bestowed by fossil fuels, a way of imagining the living collapses: the omnipotence of infinite growth, the idea of the individual human subject free from limits, and the reduction of nonhumans to "nature" at disposal and as green spaces "out there," distant from us. This reoccurs as a checkmate to the cultural imagination in the face of the "urgent proximity, intimacy and relationality of the nonhuman presence" (ibid., 5). This happens as an "emergency," where we are frightened by environmental crises and feel subjugated to other forces of the living. But this emergency might have a generative dimension: we might discover that we are necessarily interdependent, and that we have just forgotten its symbolic as well as emotional alphabet.

The very thing that is most familiar to us resurfaces as uncanny and distressing in its re-emergence, and what makes it incomprehensible is precisely our nature/culture dichotomy, with all the dimensions of denial and removal (Norgaard 2011; Weintrobe 2021).

Climate change is implicitly uncanny: weather conditions and the carbon-intensive lifestyles that are changing them, are extremely familiar and yet have now been given a new menace and uncertainty (Marshall 2014).

Between the Presence of a Crisis and the Antique Crisis of Presence

Latour (2017) in his latest texts has shown "the immensity of the ongoing catastrophe": a symbolic disorientation and blockage of social thinking, for the transmutation of values, the recognition that others act and reacts in the world in a sense of vertigo and acceleration, together with the amplifications of ever more excruciating injustices and the pervasive languages of war in the climate crisis.

Latour and Schultz (2022) ask what and how to share about the dizziness of the world swirling around us in this astonishing reversal of the links between generations in the new climate regime. It is an emptiness characterized by frenetic paralysis: paralysis, because of the un-symbolized sense of anguish; frenetic, because it follows the neoliberal productivity rhythms of increased growth and consumption in a fragmentation of time and closure of future, where the rhythms of the living and the weather show dynamics unseen even to scientists. The sense of bewilderment is linked to the failure to metabolize the effects of an epochal change that is socially traumatic, with the attached processes of denial.

As Ulrich Beck (2016) noted, it is not a revolution nor a change, but a metamorphosis of the living that requires another metaphorical outfit in order to grasp its significance. According to him, the climate crisis is an agent of metamorphosis. It has already altered our way of being in the world and shifts the focus of "being in the world." The new climate regime increasingly weighs upon all issues, upon all class relations, upon all emotions, but nothing has been done to metabolize its formidable effects. Hence the tremendous vacuum in public space. We need to turn language upside down by renaming the world "for the new-old living": names, recognitions and metaphors for shared languages in public spaces to "realign" with the "great upheaval" (Latour and Schultz 2022).

The question of naming that which remains unthinkable and unmentionable today, runs through all sciences, knowledge and arts, precisely because of the collapse of the imagery of fossil fuels; and often literature and the arts, precisely because they explore imageries, have best condensed this stalemate, as Andri Snær Magnason writes, "when a system collapses, language loses all grip on the real. Words, instead of capturing things and concepts as they should, remain in a vacuum, inapplicable. Suddenly we can no longer find terms and concepts that correspond to reality" (2020, 12).

It is precisely the collapse of Nature's foundations that shows its cosmological character, largely fossilized in the great carbon festival and the connected "great acceleration," and thus the paralysis that occurs if we lose the reference framework for saying the world, nonhumans, desires, needs, sharing and community. These are traumatic collective times because of the force with which the metabolism of the world is changing in the "great acceleration," with continuous and oscillating jolts of novelty; and because of the dimensions of violence and authoritarianism that take on legitimacy in the polycrisis, that is multiple planetary crises mesh into one another and bind to the ecological urgency, because the world faces "no single vital problem, but many vital problems, and it is this complex intersolidarity of problems, antagonisms, crises, uncontrolled processes, and the general crisis of the planet that constitutes the number one vital problem"(Morin and Kern 1993, 74).

Further, contemporary times are traumatic because we are failing to make a narrative of them, because of the difficulty of our imagination that elides into interdependencies and coexistences, even in scientific paradigms, to make them collectively meaningful and desired.

We are not in a simple disjointed seasonal transition, but in an "epochal" transition that we are striving to read with grids and models that are inevitably complicit with the fossil fuel phantasies and its myths. It is the entry into a New Climate Regime (Latour 2017) as a "great Reversal," a radical change of cosmology from the fossil one: the first step is naming the hovering collective trauma, as rituals of crisis of presence have always had to do, and thus also naming the working through of the mourning of what is no longer there: a linear future planned by development, the omnipotent dominance and human sovereignty over the living as if it were disposable, or acknowledging the reality of fragility and loss in a "damaged planet" (Haraway

2019). Trauma is the impediment to symbolization, to narratives, to tragic catharsis, to setting in motion feelings and experiences that cannot be individually worked through.

Ernesto De Martino (1962, 1977) expressed this disorientation through the notion of crisis of presence in his studies of southern Italian peasant cultures. All cultures have faced the sense of the end of the world and cultural apocalypse in their history, as well as the loss of their identity references and the fear of losing "a world" that is environmental, material and, above all, of meaning and horizons of the future. The rituals of presence involved an anticipated grieving to express and enact the possibilities of redemption from marginality, the loss of one's social worlds and worlds of meaning, and the consequent overbearing and "unthinkable" emotional dimensions. In crises of presence, however, the emphasis is on the consciousness of the crisis, as well as on the staging of invisible and nonhuman actors as a cultural and ritual fact in a shared symbolic system; it is not, therefore, a paralyzed standstill in a pre-mediatized "presence of the crisis" that we witness as a catastrophe that we must manage or evade, in a condition of repetition compulsion such as the one we find ourselves in today.

We now read reality through the perturbing presence of the crisis with certain central tropes to the collapsing imaginary of fossil cultures: individualism, where collective and social dimensions disappear, typical of neoliberal subjectivity; enchantment with matter-goods, and disenchantment about the living, which vanishes without other visions a future that flounders between success and failure of the self. Here stands the crisis as a construction of subjectivity in modernity, an aspect even more relevant in the acceleration of polycrisis, where it is difficult to understand how we process together, what we feel, how we can make the experience of these times of acceleration and continuous emergency, a shared and communicable one.

The concept of crisis, by which we compulsively interpret our present moment, has lost its generative valences of a drastic *caesura*, of mourning and regeneration, of a moment of ending but also of beginning, of impetus for change that it had assumed in antiquity. The etymological meaning of crisis, from the Greek *krino*, is to judge, to discern, to decide. These meanings are derived from a term of medico-legal origin: a moment of choice that one is called upon to make at a time of great change, a discernment that is imposed because of an alteration. We today, however, denote with this term other radically different meanings: the anxiety-inducing disorientation of the decision, the dimension of threat and insecurity in the collective indecision in the face of a closing future, or the technical-administrative and emergency "management" of the crisis.

In the modern semantics of crises, therefore, the finalistic idea of progress is already implicit -the future as man's advancement independent of the living. This idea is intimately intertwined -and we forget this- with the idea of the domination of nature and the inscription of history as a purely anthropological fact, with no other actors. An irresistible force of attraction of an openness to the future that seems to quickly infect all social strata (Roitman 2014) and that brings with it a crisis already implicitly anticipated as collapse: either abundance or scarcity, either a commodity paradise on earth or its apocalyptic opposite.

In the contemporary world, the notion of crisis has undergone a radical shift: from a choice within a prognosis, it has become a "prognosis of time" (Roitman 2014). It has secularized an apocalyptic and catastrophic meaning – or infinite growth or the end of the world- into a world of uniquely human life relations. The emphasis lies on the emotional and disorienting dimensions of the critical state, rather than on the act of choice and discernment. Crisis therefore becomes a construction of subjectivity within an idea of development that already has implied its catastrophic opposite: it already anticipates a social experience, an idea of subjectivity, a structure of the self. Gazing at reality through crisis becomes a self-referential, self-generative mode in dealing with unprecedented contingencies.

Crisis is a blind spot that enables the production of knowledge. [...] In this sense, a crisis is not a condition to be observed (loss of meaning, alienation, faulty knowledge); it is an observation that produces meaning (Roitman 2014).

We read time and reality through crises, a continuous "presence of crisis" rather than a crisis of presence as a generative symbolic process. And it becomes a "regime of subjectivity" in the common imagination. Indeed, it is first and foremost a practice of the cultural imagination, a generative attempt to make sense from meaninglessness, from the anguish and failure to give meaning to the future anymore. This happens at a time when the future had been identified in the force of an inevitable, confident, master-of-nature "time-modernity." A crisis therefore provides a framework for making a fragile sense of the feelings of inconsistency, uncertainty, instability and discontinuity, for naming a disorientation. And it shapes a social experience of chronological time, becoming a matrix that generates meanings, that reproduces crisis as time crushed in the present, where the horizon of the future is shattered.

The rituals of crisis of presence, as rituals of the end of the world, defined a generative grasp of cultures in dealing with loss, mourning and critical moments of existence, as we do today in the face of "the death of nature" and the fossil fuel-designed collapse of the future. They opened to the possibility of existential and communal redemption, the reactivation of meaningful relationships and shared imageries for change.

The crisis of presence has been defined as the bewilderment, but, in the same time, as the collective possibility of reinventing cultural forms in order to make sense of a radical change, such as the generative sharing of communities in dealing with loss and mourning, with the forms of territorial anguish, the feelings of the world collapsing. Those definitions that have historically been linked to peasant cultures at times of seasonal change and critical moments of existence, fit well into the climatic crisis: in the face of radical change there is an absence of any project, of intersubjectivity, of any capacity to go beyond; it is precisely the crisis of presence, the crisis of being in the world (Signorelli 2015).

Ideas concerning the end of the world, which abound in all cultures, were not limited to celebrating a hypothetical end in a nihilistic form, but to culturally elaborate a redeeming anguish in the face of critical moments of becoming. Referred to as cultural apocalypses, these modes of imbuing history with its radical threats have always been connected to forms of re-action of local cultures, such as apocalyptic rituals and myths, millenarian or messianic movements: rites of the future as a cultural fact, a network of often reinvented and syncretic symbolic forms to share what presents itself as irrational, as the negative of existence, to bring it into play otherwise.

Cultural rituals of the end of the world have had not had only a cathartic function, but also a function of defining and revealing limits. For example, agrarian rituals defined the very boundaries of territories and forms of common cooperative management, staging presences, dependencies and fragilities in the living and atmospheric flows, but also were performances of self-limitation, exhibiting beyond-that-human relations, changes and above all a moral community: thus, allowing negative events to be shared, to be represented in a public scene, and even to product generative conflict.

Mourning and Vegetal Masses

This is apparent in the peasant rituals, held in anticipation of the cold darkness of winter that seasonally opened at the end of the harvest in the rituals and songs of the "masses of sorrow," so well described in Southern Italy by De Martino (1962). At the end of the wheat harvest, which coincided with "the period of maximum seasonal occupation, the exaltation of vital energies, the last strenuous golden harvest at the end of many others, amid euphoria and good omen in the torrid climate, reaping ceremonies were celebrated: a masking of the action of reaping in goat-hunting in the midst of the wheat, as a device to employ the charge of aggression circulating among the reapers. (De Martino, 1962) A passage that had to be ritualized precisely because it was dangerous, to be collectively signified and processed.

As the summer cycle drew to an end with the exuberance of wheat, "the peasants experienced a critical moment of individual and collective existence" (ibid. 214, our trans.) that mourning and devoidness of plants: this represented "the encounter with the insignificant", "the experience of the plant void, a void punctuated with imponderables", the "uncertain prospect of the new harvest for the following year", precisely an experience of death lived in the wheat fields. And like any collective mourning, it masked the experience of seasonal death and the harvesting of grain, as guilt for the void created in a reparative sacrifice, with the hunting of an animal chosen from among those harmful in agriculture such as the goat. Weeping and ritual chanting, following the pattern of the traditional funeral lament, were central in accompanying this uncertain passage, a "psychological protective technique to overcome guilt" (1962, 218, our translation). and the fear of the uncertainty that opened up: "one does not know whether this void will be filled again, with the coming of the new season (1962, 218, our translation).

Rituals and songs in which not only the dimension of human uncertainty and fear was staged, but also "unfolded" that on whom it depended: it staged the entanglements with other nonhuman subjects present, with the course of the seasons with their risks of poverty and bad harvests, plant diseases, epidemics. In short, it made manifest the profound interdependence and co-fragility, because interrelated to other actors and forces. And De Martino's concluding notes are interesting, taking us directly back to us: these "cereal societies ignored the relative security that proceeds from the modern experience of production by machines which falls entirely under human control" (ibid., 219).

Today we find that the relative security of farming amid environmental changes has never been under human control, on the contrary we are in a dynamic that is "out of control" (Eriksen 2016).

Even the seasons and their passage, are bewildering and continually jeopardize the harvest, producing a sense of "emptiness." Besides, the environment is not "under human control," because we have stripped ourselves of the social and symbolic textures to working through griefs and traumas. And if we do not name them, as an ancient and still fertile history tells us, we pretend they are not there, but we remain terrified.

Thus, not objects of "nature," but scenes of man's exposure to the forces of the earth, to its presences on which one depends, of the always uncertain and risky seasonal transitions: communities and their rituals, prayers and chants did not leave out mourning, separation, grief, which they rather took by the hand. Today not only have we stopped singing the praises of these actors which are so important to giving meaning to the passing of time, cycles of life and death, but the cycles themselves are in fluctuation, they give no order: they do return, but not as we expected them.

And that is precisely what we are in want of today: a consciousness of crisis, a resignifying of this moment of choice and discernment as a collective dimension. What is missing today are the cultural coordinates, the models for reading change, which make the climate crisis a public silence, rather than a public space of climate citizenship.

A fluctuation between fullness and emptiness characterizes our perception of inhabiting these times of crisis: a fullness of modernity's narratives of the future in neoliberalism, an emptiness of their ability to make sense of what cultures experience in local dimensions, amid marginalization, estrangement, and disorientation: where we fail to name and shared systems of meaning to what we experience.

It is no coincidence that generational movements take on collapsological names, related also to scientific definitions of ecological collapse: "Extinction Rebellion," "Last Generation," express community belonging and action on a stage that is situated in the climate crisis, in new rituals of crisis of presence, emergent and contradictory also; where "nature" however becomes political and acts, where they publicly dramatize the epochal ecological dimension with the privileges and negations of the fossil fuel system in a new generational social alphabet.

Uncanny Natures and Desires of Environmental Relations

It is not only nature as a separate, "at disposal" (Rosa 2018) mute, passive field that has never existed, but it is the cosmological foundations of Modernity's naturalism that have collapsed: "We need to diagnose the source of this paralysis and to seek a new alignment between anguish, collective action, ideals, and the meaning of history" (Latour 2015, 53).

In the face of paralyzing and individualized anguish, in the face of knowledge disciplined by Nature and the economic, undisciplined knowledge seeks new paths and alliances in feeling subjugated by "others" removed from view: other agents who remerge in beauty and in a sense of reality. We live in a "framework that now reacts to our actions, viruses, climate, humus, forest, insects, microbes, oceans and rivers which opens to an extension of sensibility"(ibid, 53), even with scandal: it represents a change in cosmology, a redistribution of no longer only human capacity to act and manage, and of a widening of subjectivity where we had subtracted it, which can also open to collective action (Latour 2015).

Discovering, in the Anthropocene, that we are interdependent with bonds of co-fragility and entanglement within the living is undoubtedly a real earthquake erupting from the atmosphere, shattering the certainties and many declarations of independence of Modernity. "The whole world has already changed the paradigm" in the dual sense that the living subjugates us, but also that so many communities, collectives and experts already think themselves in this epochal change.

We are witnessing a new climatic regime that overturns maps and points that have magnetized and attracted Moderns. This requires a broad effort to situate ourselves in a place that we should attempt to describe with others, a search for shared maps and metaphors in knowledge, science, and common sense to understand where we are and where to go. (Latour and Schultz 2022)

The reversal leads to the sad passions precisely because nature turns out as uncanny and everything we had stabilized as available resources, natural capitals, things of the world as reassuring, continually and acceleratingly re-emerge as unstable, unpredictable, active, unknown, uncanny in a continuous and frenetic game of denial, concealment and re-emergence (Van Aken 2020).

According to Latour and Schultz (2022), the world of the nature of the economy simply it does not exist.

Most cultures have not imagined their relationship with environmental actors, in producing food, in praying, in sociality, in politics, in a separate and distant field called by us "nature," without subjects, relationships to our history. What we view at a distance as nature is named elsewhere through parental, political or sacral terms, where even selectively and contextually, the elements of our "nature" is subjectified. And this is not only in cosmologies or symbolic constructs, but all the more so in resource management systems, irrigation relations and practices, and landscape constructions, thus in the everyday practices of making food, habitat or reproducing community (Van Aken 2012).

Naturalism is thus a cosmology of ours: society is not thought of in an environment but frees itself from it *as if* it was "outside."

Philippe Descola (2005) identifies alongside our naturalism, three other main forms of socializing the environment: totemism, animism, and analogism. Each of these major families, encompassing quite different cultures and environments, is always a defining model of the human being and his relations in connection with other ecological presences, they are forms of "identification" in relation to, and in dependence on, "others." They are also patterns of ethos, i.e., of moral, value, legal systems of humans in interdependence with other beings or forces of the living; and they are not 'closer to nature', since there is no opposite field to approach as such, but distinct, acting and related presences, part of our social world.

On the other hand, the prevalence of the oppositional nature paradigm in naturalism, implies that an interface between humans and non-humans is denied, the common entanglement: hence the fear when we rediscover that which we have imagined as objects available in the most varied forms resurfaces as a great swarm of emergent, and "unavailable" subjects (Rosa 2018). In naturalism, the environment is characterized as a largely passive object to human action, not a subject of relationships and history, of action and retro-action.

Moreover, nature is thought of as "being at the disposal of humans," whether for scientific exploration or for intensive exploitation or environmental conservation in clearly demarcated green enclosures, where an idea of technological mastery over that which is not human prevails at the base. This is indeed a grand declaration of independence, in today's increasingly intense interdependencies, for example in a metropolis: a Promethean notion of man as a creator, which renders an identifying value to the avoidance of limits. However, this leads to the removal of the finiteness and fragility of the human, a crucial element in all local and economic knowledge of cultures, and the idealization of the unlimited and of "freedom" without alleyways, typical today of neoliberal rhetoric.

Therefore, subjects reemerge today that we no longer know how to connote: the etymology of environment, derives from *environing*, "that which surrounds us," that in which we are co-involved and co-entangled: not dumb objects but subjects of enveloping forms of life, just like what circulates in the atmospheric environment we inhabit. (Van Aken 2020) The problem is not so much that "nature" changes in climate systems but that this change is a checkmate to the idea of human mastery, and we no longer have the words to talk about relationships and affect with the life forms on which we depend. The language of science and common sense in addressing the environmental crisis thrives on metaphors, and a "tragic absence of environmental frames" (Lakoff 2010).

Thanks to the energy-generating power of fossil fuels, we have constructed an idea of the human as being outside, distant and independent from the environment; sheltered in a climate-controlled, indoor inside, we awaken today from this dream turned into a nightmare. In the carbon economy, we have altered the subjects of the environment to which we are interrelated, as mute objects, out of relationships: objects that today paw, vibrate, pulsate and frighten us. In fact, there has never existed, except in the imaginary construction of fossil fuel modernity, a nature that is thought of as an opposite, as brute matter, a great other: nature as a warehouse, as a dump, or as a grand spectacle as it has staged in modernity. It is time to mourn this cosmology, in order to be able to generate other, to be able to "return to earth" (Latour 2017) in the new climatic regime.

Latour draws a comparison of co-present planets imageries at the centre of local and global politics of nature today on this scene. In the first map, there is a "globalization planet" (Latour and Schultz 2022) that continues to attract those who seriously hope to modernize as they once did, regardless of the gradual disappearance of the earth on which they live. To be human is to remain blithely indifferent to the fate of the planet. In the second map there is a "Planet Exit," inhabited by those who have understood all too well the limits of the earth but for that reason have decided to abandon it, at least virtually by inventing hypermodern bunkers. For them, the word human being is reserved only for the rich. In the third map there is a "Planet Security": that of the outcasts who group themselves into solidly confined nations, themselves completely de-territorialized but which they hope will protect them by erecting more and more borders and walls.

Last but not least, a majority of Earthbounds, those who are bound to and belonging to the earth, who agree to inhabit on earth grasping what allows living beings to make the earth habitable. While the figure of "returning to the earth" is central, we cannot forget that the earth from which we have tried to take off is no longer the same in a context of our ignorance of what means to inhabit the earth. A return therefore takes place among the ruins and with new interactions that nevertheless allows us to realign the "world we live in" as much as possible with the "world we live of," the world of "resources" elsewhere, invisible, finite, in other territories. Latour calls them "menders" who must strive to recreate another weaving of those territories that their enemies have abandoned after plundering them, as social movements, activists, ordinary citizens have been doing for decades in buying groups, in changes in consumption, in socializing otherwise electricity and its sources, in defending local territories and waters, in short in decarbonizing the imaginary from everyday and shared practices (Van Aken 2021).

Pervasive Petrocultures

Our naturalism has been estranged by dichotomously demarcating what remains most intimate and familiar to us, our interdependence with the organisms and networks of life with which we live; and this even more so in metropolitan dimensions where we are actually even more interdependent with environmental networks, also referred to as socio-natural networks (Van Aken 2012).

For what is missing today are precisely those cultural coordinates, those models for reading changes and processing our feelings: it is impossible to accept, process, even love interdependencies with sovereignist models of the environment! It shall therefore be necessary to process the grieving of the fossil fuel cosmology in order to be able to generate another, as so many families, communities and territorial experiences have been doing for a long time in the way of inhabiting, caring, and being cared for, by territories, in a conversion of desires; and in a letting go, and passing into the past, the attachments and desires of oil cultures (Fletcher 2018).

The system of production has become a system of destruction of the planet's conditions of habitability with the brutal return of planetary limits: the very secular salvation of the commodity paradise on earth, of consumerism as a globalized social ritual practice, shows its destructive dynamics. This is not only a process of intellectual understanding: the reversal is also a "realigning" of the magnet of the affects and senses, of our enchantments and desires, beginning with the "disanimation" of commodities at the center of our cosmology and habitus; that "misplaced immanentization"(Latour 2015, or problematic materialism for so much literature in the Anthropocene, connected to the deification of the human being, his dreams of immortality and independence, its making itself an "exception" to the living, which opens to central themes in the social sciences today, such as the lack of social alphabet in the atmosphere and the secularization of the sky turned into socially insignificant logistical space (Van Aken 2020).

The naturalistic myth of a world out there at our disposal and of the exceptionalism of the human are deeply connected to the fossil and the cultural imagination it has captured: it is not the ecological end of the world, but certainly the end of a way of representing the world, a shift in paradigms and perceptions of place in the living.

Our entire economic and cultural system is deeply intertwined with the social worldview of fossil fuels: ideas of man as outside the environment, desires for things and consumption, and ideas of freedom and well-being. It is not only geological reservoirs of the fossil fuels that are being depleted, it is the fossil reservoirs of meaning that are being depleted by their destructive dimensions of the forms of habitability of the next generations. It is their utopian and cultural dimensions that need to be demythologized: black gold deposits have been mythical reservoirs of hope, symbolic forms that have redefined what is desirable, dreams (the American dream), rites of passage, collective unconscious fantasies, into an effective omnipresence, ubiquity and dependence not only material but emotional and social. A mystical dimension of fossil power has accompanied their social invisibility in a "petroleum magic realism" (McDermott Hughes 2017) that has saturated the thinkability of modernity as a spectacle of power, well-being, and environmental control, so that other dimensions of the future no longer seem imaginable.

Indeed, the dreams of fossil fuels has fed the idea of a commodity paradise on earth, where nature could be imagined as cheap and infinite, becoming an inanimate support and commodities themselves take on new souls, values, and beliefs. The fossil fuel is the basis therefore of a moral, value, identity, and ethical system of ideas of the "human" and the individual.

The absence of fossil fuels in social analyses is closely related to their pervasiveness, since it is what has nourished the ideas of "nature" as a field separate from the human, that are so pervasive in scientific paradigms and in the dichotomy between the natural and human sciences; it "invented" the myth of infinite growth in a finite world; it transmuted the ancient foundational value of limit into valorization of the unlimited as *hybris* of omnipotence, reinforcing the idea of a sovereign, individual and self-sufficient human subject.

It is this forging of a symbolic apparatus and fossil cosmology that has ambiguously captured our imagination: on the one hand, they are held out as icons of promise and ideas of messianic horizon and secular salvation, on the other hand, they have for decades, been the foundation of unusual privileges over growing poverty, but also of a dimension of potential ecological and climate-collapsing future conditions of habitation.

This cultural invisibility of fossil fuels has been a ceaseless process of denial and masking: even today where the correlation between fossil-based productions and climate-changing gases has been proven, fossil fuel as an assemblage of actors among fossil fuel multinationals, banks, financial and insurance investment funds, the military apparatus to protect it, remains hidden, and these same actors promote green sustainability policies and are subsidized and idealized in global politics: Oil is part of our lives in a fundamental way, permeating our cultures, economies, politics and material lives (Åberg, Ekberg, and Lidström 2023).

"Oil networks redefined forms of power based on energy control even in extractive contexts" (Mitchell 2011, 120): despite coal requiring mining, miners, logistics, and transportation, and have allowed for the ability of labor movements to strike and stop production to obviate their ability to demand rights, oil managed to overcome those political rights. The carbon economy enables and induces oligarchic, deeply centralized networks with a direct proximity to forms of authoritarianism; it has founded ideas of politics and "carbon democracy" (Mitchell 2011). Moreover, its "discovery" as a mechanical fuel and its centralization within imperial politics allowed for the construction of an idea of universal modernity identified with the West as singularity and uniqueness.

Fossil fuel cultures lie at the center of the very symbolic production of modernity, structuring ideas of subjectivity and desires, effects and fantasies, an aspect that makes it so tortuous, but yet fertile, to decarbonize the cultural imagination. What has been explicit for some time in social movements and community making has been the reconversion to other desires of environmental relations, such as mutual care, given that one is cared for in healthy environments; and in the critics of the fossil fuel style of life, other ideas of limits such as foundations, of ecological interdependencies with nonhuman subjects, have opened up other senses of generative futures. As Latour and Schultz (2022) write, it is not just about processing the effects of climate change, but the "affects" toward what one depends on and is "attached to."

"It was the transformation of nonhuman beings into dumb resources that enabled the conceptual leap as a result of which it became possible to reduce the earth and all it contained to inertia" (Ghosh 2021, 45), a transformation that finds the fossil as its own actor, a mythicization of modernity that makes humans appear triumphantly free of material dependence on the planet; a sense of omnipotence that has lost the social and aesthetic alphabet of interdependence and co-vulnerability in the living, and hence the impediment to symbolization that becomes unprocessed trauma.

"It's not climate change, it's everything change" (Atwood 2015): uncertainty and risk emerge in the very forms of belonging to which we have delegated dimensions of salvation, universalism, exceptionalism, and the future in an escape from reality. The material and symbolic dependence on the fossil resembles a drug addiction whose abstinence cannot be sustained: the carbon grand festival and the carbon economy continue their inertia with scenarios of ecological collapse that carbon itself reproduces in a dynamic of habituation and anesthetization. Precisely because of the mythical dimension of fossils, one persists in subsidizing and idealizing this idea of humanity and modernity, going on to perpetuate climate-altering elites who have not only been funding climate denialism for decades, but hiding the incredible inequalities that environmental changes cause.

Energivorous Freedoms

Ryszard Kapuscinski, as early as 1982 emphasized these magical and imagination-capturing aspects. According to him, oil kindles extraordinary emotions and hopes, since oil is above all a great temptation. It is the temptation of ease, wealth, strength, fortune power. Oil is a fairy tale, and like every fairy tale, a bit of a lie.

It has also been called energopower or "oil culture," where modern capitalist society comes to coincide with an oil society at its heart and rampant consumption has become an archetype through its culture. (Boyer 2014) A petroptimism that reproduces in the policies and rhetoric of green neoliberalism the same cosmology of human and nature available as background.

"No commodity has caused so much damage while provoking so little antagonism" (McDermott Hughes 2012, 1) a unique combination of hopes for emancipation and dystopias: a mythical dimension has also concealed criticism, contestation, and conflict that have been far from absent (Pessis, Topçu, and Bonneuil 2013).

As Cara Daggett (2018, 27) shows, coal and oil do more than ensure profit and fuel consumption-heavy lifestyles. If people cling so tenaciously to fossil fuels, even to the point of embarking upon authoritarianism, it is because fossil fuels also secure cultural meaning and political subjectivities and fossil fuels matter to new authoritarian movements in the West because of profits and consumer lifestyles, but also because privileged subjectivities are oil-soaked.

Indeed, oil has redefined ideas of subjectivity, and in particular of individualism as autonomous from the world, where freedom comes to coincide with energy-driven practices and needs, with desires of the sovereign individual subject. Since 1800 the population has increased seven-fold, and energy production 28-fold, an element that well highlights the dependence of the modern subject and its "needs" on high energy consumption "as if" it was outside the living. The myths of infiniteness and limitlessness, that "playing God" (Eriksen 2016) from which we awaken today in our incompleteness, fragility and dependence, derive precisely from those fossil dreams of immortality. That we can compost in the past.

Conclusions. Decarbonizing the Cultural Imagination

At a national rite, a secular mourning, called for the death of the Ok glacier in Iceland, a "letter to the future" was written. Already declared a dead glacier by geologists since 2015, in 2019 the Icelandic state decided to celebrate an actual rite of presence for a subject that, though nonhuman, with its loss revealed the end of a central identity, affective, emotional presence, as well as resource. A first mourning of loss among 226 others glaciers that are expected to disappear by the end of the century. The inscription reads as follow:

A Letter to the Future
    Ok is the first Icelandic glacier to lose its status as a glacier. In the next 200 years, all our glaciers are expected to follow the same path. This

monument is to acknowledge that we know what is happening and know what needs to be done. Only you know if we did it.
August 2019-415ppm CO_2

What is crucial here is the change in imagery: a glacier emerges as a "person" in its loss; policy takes on ethical and political responsibility for the change in world references, above all, of the future: today's policies define or undermine the "futurability" of the "newcomers," as Hannah Arendt called them in another era. The national discourse no longer denies the paradigm shift, rather, it accommodates the social and cultural dimensions central to the climate crisis, those overbearing emotions and disorientations that are "enacted" rather than denied or habituated in a managerial rhetoric and technical reductionism. Most importantly, the definition of time (past, present, future) changes: the date is correlated with the average measure of global CO_2 emissions: the social alphabet of fossil and carbon becomes visible, takes center stage, and imposes itself to talk about the future.

Despite "we know what is happening and what needs to be done," despite the fact that renewables are cheaper than fossil fuels and can clear the way to decentralized and reterritorialized patterns of energy, and consequently, of decentralized political citizenship; despite the dimensions of ecological collapse and degradation and impact on forms of local autonomy, steps are slow and contradictory and aspirations filled with uncertainty.

A decarbonization of the economy cannot be divorced from a decarbonization of the cultural imagination and a redefinition of the desires of what makes up a good life, toward connected forms of energy sobriety and a cooling of our needs and ideas of collectivity in the living. What had been completely delegated to technical management, may now return, although in conflict, owing to a deep resistance and delay, to being social, to being participatory, decentralized, political, aesthetic.

As Margaret Atwood (2015) has well summarized, there is an intimate connection between energy sources, desires, and constructions of the self:

> "in the energy culture of coal, a culture of workers and production, in which 'you are your own work', typically defines working-class culture and the possibility of claiming oneself as a class in solidarity within the production system. In an oil and gas culture -a culture of consumption- you are what you own, 'I am what I buy'" are conditions that open to cultures of consumption, endless growth, and enjoyment of material goods. (…) In a culture of renewables, 'you are what you keep, 'I am what I save and protect'" (Atwood 2015)
>
> These represent the only generative form to "return to the earth" as a sense of reality and open shared aspirations and redemption of the future. And this puts back at the centre the need for a reorientation of

emotional dynamics and desires: in a new, but very antique at the same time, "freedom to depend." (Latour and Schultz 2022)

These dynamics highlight the bonds and limits that liberate, and enable, and are already enabling from below, the ability to say "I can act like this." After the historic overturning of the notion of limit in Western thought and common sense, with its removal and coincidence with deficit, limiting oneself can become a foundational desire that opens up a revaluation of the notion of freedom itself. Redirecting the aspirations, fears, and desires associated with freedom, which have always changed throughout history, occurs through the extension of sensibility to multiple actors, but especially to multiple relations and interdependencies, an immense expansion of sensitivity to the conditions necessary for life.

It is a discovery of bonds that make us free, a release from negative conceptions of freedom as an escape from constraints and an escape from reality, to a historical positive connotation of that which enables communities to live together autonomously in an ongoing apprenticeship of dependence. The more we depend, the better. And this is what is often happening from local movements and associations, social networks, forms of activism and thinking together, because people have already realized that they have changed worlds and that they live in another land, in an attempt to slow down in order to identify alliances to build (Latour and Schultz 2022).

For it is time itself that takes on new value and new scales, when we realize that we have already changed climate systems for the next centuries and millennia, but we may not swing and tip too much into a radically different new climate regime.

Relationships between generations have radically changed, in an unprecedented intergenerational transmission: we pass on debts and potential for uninhabitability to future generations through ignorance, cynicism, sloth, privilege, and fears of the present. But as Magnason (2020, 26) writes, in this framework playing out other worlds to be transmitted and changed, with the generative dimensions of hope with the abandonment of fossil mythologies, takes on new and dense meaning:

"Think, 262 years. That's your span of time. You know people who cover it all. Your time is the time of someone you know, love and influence. And your time is also the time of someone you will know and love, the time you create. You can have a direct influence on a future as long as 262 years. Grandmother teaches you, you will teach your great-granddaughter. You can affect the future until 2186."

References

Åberg, A.Ekberg, K., & Lidström, S. (2023). "Pervasive petrocultures: Histories, ideas and practices of fossil fuels". *Journal of Energy History, Revue d'histoire de l'énergie*, 10.

Atwood, M. (2015). "It's not climate change, it's everything change". *Medium.com/ Matter, July* 27. https://medium. com/matter/it-s-not-climate-change-it-s-ever ything-change-8fd9aa671804.

Beck, U. (2016). *The Metamorphosis of the World. How Climate Change is Transforming Our Concept of the World*. Cambridge, UK: Polity Press.

Boyer, D. (2014) "Energopower: An Introduction". *Anthropological Quarterly*, 87(2): 309–333.

Daggett, C. (2018). "Petro-masculinity: Fossil fuels and authoritarian desire". *Millennium: Journal of International Studies*, 47(1): 25–44.

De Martino, E. (1962). *Furore Simbolo Valore*. Milan: Feltrinelli.

De Martino, E. (1977). *The End of the World: Cultural Apocalypse and Transcendence*. D. L. Zinn (Trans). Chicago, IL: The University of Chicago Press, 2023.

Descola, Ph. (2005). *Beyond Nature and Culture*. J. Lloyd (Trans). Chicago, IL: University of Chicago Press, 2014.

Eriksen, T.H. (2016). *Overheating. An Anthropology of Accelerate Change*. London: Pluto Press.

Fletcher, R. (2018). "Beyond the end of the world. Breaking attachment to a dying planet". In I. Kapoor (Ed), *Psychoanalysis and the Global* (pp. 48–69). Lincoln, NB: University of Nebraska Press.

Ghosh, A. (2016). *The Great Derangement. Climate Change and the Unthinkable*. Chicago, IL: University of Chicago Press.

Ghosh, A. (2021). *The Nutmeg Curse. Parables for a Planet in Crisis*. Chicago, IL: University of Chicago Press.

Haraway, D. (2016). *Staying with the Trouble: Making Kin in the Chthulucene*. Durham, NC: Duke University Press.

Kapuscinski, R. (1982). *Shah of Shahs*. San Diego, CA: Harcourt Brace Jovanovich, 1985.

Lakoff, G. (2010). "Why it matters how we frame the environment". *Environmental Communication*, 4(1): 70–81.

Latour, B. (2015). *Facing Gaia: Eight Lectures on the New Climatic Regime*. C. Porter (Trans). Cambridge, UK: Polity Press, 2017.

Latour, B. (2017). *Down to Earth. Politics in the New Climatic Regime*. C. Porter (Trans). Cambridge, UK: Polity Press, 2018.

Latour, B. & Schultz, N. (2022). *On the Emergence of an Ecological Class: A Memo*. J. Rose (Trans). Cambridge, UK: Polity Press, 2022.

Magnason, A.S. (2020). *On Time and Water: A History of Our Future*. L. Smith (Trans). London: Profile Books, 2021.

Marshall, G. (2014). *Don't Even Think About It: Why Our Brains Are Wired to Ignore Climate Change*. New York: Bloomsbury.

McDermott Hughes, D. (2012). *Paradise without Labour: How Oil Missed Its Utopian Moment. Piscataway, New Jersey*: Rutgers University.

McDermott Hughes, D. (2017). *Energy without Conscience. Oil, Climate Change, and Complicity*. Durham, NC: Duke University Press.

Mitchell, T. (2011). *Carbon Democracy: Political Power in the Age of Oil*. London: Verso.

Morin, E. & Kern, A.B. (1993). *Homeland Earth: A Manifesto for the New Millennium*. New York: Hampton Press, 1999.

Norgaard, K.M. (2011). *Living in Denial. Climate Change, Emotions and Everyday Life*. Cambridge, MA: MIT Press.

Pessis, C., Topçu, S., & Bonneuil, C. (2013). *Une Autre Histoire des "Trente Glorieuses". Modernisation, Contestations et Pollutions dans la France d'Après-Guerre*. Paris: La Découverte.

Roitman, J. (2014). *Anti-Crisis*. Durham, NC: Duke University Press.

Rosa, H. (2018). *The Uncontrollability of the World*. J. Wagner (Trans). Cambridge, UK: Polity Press, 2020.

Signorelli, A. (2015). *Ernesto De Martino. Teoria Antropologica e Metodologia della Ricerca*. Rome: L'Asino d'Oro.

Van Aken, M. (2012). *La Diversità delle Acque. Antropologia di un Bene Molto Comune*. Lungavilla: Altravista.

Van Aken, M. (2020). *Campati per Aria*. Milan: Elèuthera.

Van Aken, M. (2021). "Decarbonizzare l'immaginario culturale". In M. Acanfora & G. Ruggieri (Eds), *Che Cos'è la Transizione Ecologica. Clima, Ambiente, Disuguaglianze Sociali* (pp. 173–179). San Giuliano Milanese: Altraeconomia.

Weintrobe, S. (2021). *Psychological Roots of the Climate Crisis. Neoliberal Exceptionalism and the Culture of Uncare*. London: Bloomsbury.

Chapter 5

Illusory Immunity and Actual Inhumanity

Ronny Jaffè

Introduction

Referring to the social writings of Freud, to Bion (1961), to Kaës (2012), and to Italian authors such as Corrao (1998), Neri (1995), and Gaburri (2014) on group dynamics, I will focus on the concept of illusory-delusional immunity as regards the Covid-19 pandemic and more generally environmental changes. Illusory-delusional immunity can be associated with the inhumane cruelty towards others and nature, creating states of panic, violence, indifference, and the banality of evil.

Freud argued that humankind must beware of the myth of progress and the idealization of technology.

Contemporary society is stuck in the firm belief that absolute tragedies occur only when provoked by wars and hence governments, since the Second World War, have kept trying to cover the risks and cynically displace these conflicts in areas of the globe mainly characterized by poverty, where humanitarian catastrophes are prominent; this disdainful and blind politics of dislocation of conflict is been magnified in Europe since the outbreak of the Russian–Ukrainian war, and since late 2023, since the war in the Middle East; when rulers act according to the "fight-flight basic assumption" (Bion, 1961), thus inciting groups and masses towards hatred against each another, or act according to "dependent basic assumption" (ibid.) idealizing omnipotent technology, they become both blind and deaf when facing the different natural phenomena that may turn into actual catastrophes created by humankind.

This narcissistic destructive position reveals the human blindness of not considering that nature has an extraordinary vitality and power that, if not taken care of nor respected, can undo the progress made by civilization that conferred dignity on human beings through cultural, social, and economic emancipation after the 17th-century revolutions.

The limits of human beings, which are obscured by collective illusory-delusional fantasies of technological and scientific omniscience, transform themselves into an omnipotent and un-limited ideology, with all the risks

DOI: 10.4324/9781003498605-5

that follow. Being able to introject the sense of boundaries and tolerate the reality principle as far as the habitat in which we live is concerned, may allow us to behave in ways that are compatible with sustainability, both by freeing ourselves of these false illusions and by accepting the dark reality where, in this case, we passively accept Cassandra's prophecy – the seer who tells the truth but is never believed – to be a cruel fact, which allows this generation to do whatever it wants without giving any thoughts to its descendants.

The pandemic, the Russian–Ukrainian war, and the Middle East conflict, along with ecological damage, have created a tangled web of catastrophes since the beginning of the new millennium: the virus and the war hit the individual and globalized society, taking into account that the ideas and behaviors of the majority of people are shaped by the political and social environments in which they are created: implicit here is the narcissism-socialism polarity (Bion 1992); if the former tendency prevails, the sharing attitude of the community will break down and dissolve; a sufficiently healthy habitat should contain individuality areas, community areas, and transit areas between the individual and the group, thus encouraging the transformation from the split between "socialism and narcissism" to the dynamic cohabitation between these two polarities.

In the early days of the pandemic, it was believed that the virus made us more aware and sensitive to ecological and environmental themes. However, it is illusory/delusional to think that the pandemic may be perceived as a therapeutic strategy to heal a planet that has been abused for too long (Maffettone 2020), with the unconscious aim of giving paradoxically once again space to a voracious and cannibalistic drive for appropriation: a common behavior is to believe ourselves immune from all the severe situations imposed by nature, under the illusion that nature is immobile and silent, despite the warmongering inhumanity that overwhelms humans and the environment.

In *The Magic Mountain* (1924), Thomas Mann wrote that disease is conceived as something far from being noble and venerable, and on the contrary is considered a painful humiliation for a human being.

Mann's thoughts have taken me back to the early stages of the lockdown, when the many people who were falling ill would, for some time, keep their illness secret, as if it were a shame, a disgraceful humiliation: very probably keeping a secret is the ancient legacy of the personification of the plague victim. However, it is also connected with a primitive and archaic identification, with the mass tending to look the other way and exalt, as at the present time, a health obsession that disregards illness and psychic suffering. It has to do with a denial of reality characterized by maniacal defenses to avoid consciously and unconsciously the pain and the anguish for the wounds and the chasms spreading like wildfire on our planet. Whereas split and denial can have a legitimate defensive meaning in the individual in relation to the memory of traumas and personal suffering, it can become dangerous and perverse when it is knowingly used by the "meta-social guarantors" (Kaës 2009), because it hampers the imperative to take responsibility for the other and to take measures and remedies to take care of the group and the environment in which we live.

Split and denial have to do with the tendency to oblivion that, according to Paul Ricoeur (2000) is articulated between inexorable forgetting and the forgetting of the immemorial, an oblivion that, in addition to preventing the recall of memories erases the matrix, the very origin of remembrance. It is a significant simile between the present time and the beginning of the 20th century.

There has been an almost complete amnesia about the Spanish epidemic which produced more casualties than the First World War; it has been a mass repression which confined the Spanish epidemic in the shadow and almost no reference has been made of it in history books and in many other documents. By 2025, five years after the onset of the pandemic, Covid seems to have been forgotten and replaced by the Russian–Ukrainian war and Middle East conflict, especially in the spectacularization produced by images: mass media (television and social networks) and governments tend to penetrate our intimate and private lives, through propaganda, consumerism, and abuse of power, with apocalyptic scenarios that emphasize, distort, and disguise reality in its cruel and painful truth. It is as if mass media and governments want to create a virulent spectacular illusion to attract as many people as possible; the task is to give a (pseudo)realistic effect which is more powerful than reality; in this way, reality becomes insignificant and inessential, with the result of flattening and deleting the capability of thought and of opinion of the subject and of the group.

Spectacularization of Pain and Immunity from Suffering

"Leontius, son of Aglaion, [...] noticed some corpses lying at the executioner's feet. And he felt a desire to see, but at the same time he could not tolerate that spectacle and averted his eyes from it: For a little while he struggled with himself and covered his eyes, then overcome by desire he opened them wide and rushed near the corpses."

Plato, *The Republic*

An extraordinary book by Susan Sontag, *Regarding the Pain of Others* (2003), highlights, among other aspects, that television news gives bloody news flashes around the clock in the headline, referring to wars and other acts of brutal violence; Sontag's reflections are quite suitable also for appropriate considerations about how television, social media, and newspapers have put the suffering of Covid patients on front and centre; until recently, people suffering from Covid were shown taking their last breath, their bodies emaciated, destroyed, bloodless, in nursing homes for the elderly, hospitals, and tent cities; now, sick people have been seamlessly replaced by the bodies of the civilians and the wounded or dead soldiers lying in the streets, in hospitals, or in mass graves.

These pervasive and perpetual images on the screen, an endless flow of home entertainment provided by the media, turn all viewers, whether they want it or not, into voyeurs of macabre, almost pornographic images, also because, as Burke argues in A *Philosophical Enquiry into the Sublime and Beautiful* (1757), there is someone who experiences a certain, and not small, delight in the real misfortunes and suffering of the others; there are those who are pervaded with the *vulnus* of primitive identifications of being in the place of someone who is ill, wounded, or dead, thus creating states of nameless anxiety; the mass invasion with macabre and flaunted images, regardless of the fragility of the other, does not allow for the necessary separation between someone who is observing the tragedies in progress and someone who is under the falling bombs; the painful sharing and caring for the other can turn into a withdrawal into oneself by failing to offer one's personal, ethical, or social contribution.

However, I believe that, generally, most people can become addicted, or rather, immune to the repeated exposure to these images, with the result that pain and emotions, as far as we can see, can disappear with a gradual acritical adaptation to the situation leading to a denial of the complex surrounding reality.

Media have increasingly had the power to intercept the range of emotions and affections of the individual and of the group, thus transforming painful human events, like the ones we are witnessing, into an inhumanity with tragic implications, and Sontag's proposal of "an ecology of the images" is at risk of remaining misleading and of becoming accessory to the tragedy one is personally experiencing.

When referring to the continuous flow of these images, I do not think that they can be compared with those photographs and documents that had and have the purpose of bearing witness to the catastrophes and evils committed in the past and in the present (let's think of Frank Capra, Margaret Bourke-White, or David Attenborough, among others); documents that go deep inside, touch your heart, and not images that remain on the surface, wearing you out with boredom, or feeding a voyeuristic compulsion to repeat.

This aspect inevitably, and exponentially, increased during the pandemic, when the subject had to withdraw into himself and turn to the media world, thus miserably ending up being an ego reflected in a mirror (Lévy 2020) with a sense of self-degradation, which in time has been at risk of failing to be alleviated by the presence of the other. Covid is among those diseases which can be perceived not only as lethal but also dehumanizing. Unfortunately, isolation has proved a necessary and inevitable condition, but, in addition to that, a further damage has been produced by the so-called social and political guarantors: they have created a slogan rooted in a presumed wisdom "inviting us to travel in our own rooms" (ibid.) between the pleasures of browsing telematic media (from the sublime of music and art to the decaying into endless games and pornography) and the duties of remote working, already arguing that this would become a solution to be largely used also in the future, when the Covid pandemic would end, without duly considering how the prolonged deprivation of the group and of sensoriality could become extremely harmful for the mental health of the persons. Technology has somehow replaced the natural dimension consisting also in the individual's motivation to stay with the others, in contact with and close to each other's bodies, senses, gestures, and also with those words that cannot often be mediated by the screen.

Humankind and Nature

In the now distant 1930, in *Civilization and its Discontents*, Freud argued that although they had attained a high level of civilization, human beings had to beware of the absolute myth of progress, of the glorification of technology, of its use for eminently utilitarian purposes; besides, it was essential to protect oneself from dreadful and powerful Nature; in *The Future of an Illusion* (1927, 6), he warned that "human creations are easily destroyed, and the science and technology that have created them can also be used for their annihilation" (Schinaia, in Malidelis 2019).

These considerations by Freud on the relationship between the civilized human being and nature have uncanny connections with what occurs within our bodies and specifically in our immune system: in the external and environmental reality, the non-compliance of the environment and consequent unscrupulous violations and actions of human beings on the climate and their complex productions and creations (from global warming to pollution, to urbanization and hyperbolic cementification, to deforestation) can alter the natural elements that do not always resist and that can violently impact on what humankind, feeling omnipotent, has contrived to invent and challenge by pushing the boundaries and failing to respect Mother Earth. We can find an evocative consonance with the immune system of our body, which can sometimes have uncontrolled and paradoxical reactions. Confronted with external agents, the immune system can "create an excess of defense in the body, which in an effort to beat the enemy damages itself because of cytokines storms that cannot be defeated" (Ferro and Dionigi 2020, 90); the immune system thus betrays the body it should protect, and we have observed how "Coronavirus has disrupted the rules of immunity" (ibid.) The concurrent cause, Dionigi and Ferro add, is the defeat of the ecosystem's natural defenses, which probably promote the proliferation of new viruses (ibid.); this implies that viruses can also spill over: this is a process where a pathogenetic element of an animal develops and can infect, reproduce itself, and be transmitted in a human being; this infection is called zoonosis. As far as

Covid-19 is concerned, the most accredited hypothesis is that the virus has been transmitted from a bat to a man causing a mutation inside the organs of his body.

All this requires a radical change in the environmental as well as social and health policies in a globalized world constantly hit by viruses and their hybrids.

Since humankind is part of nature, and it does not only mean culture, progress, science, technology, the uncontrolled, nonsensical, not-thought aggression against the body and in general at the nature as regards the present and the future, becomes self-aggression; the paradox is that scientific, technological, and post-industrial hyper-modernity, supported by the arrogance and incompetence of technocracies, dictatorships, but also democracies, has neither managed to protect the ecosystems nor to create its own immune defenses, with the result of breeding hybrid little monsters; in this respect, we can think that science fiction – Cassandra was right! – is more advanced and more realistic than science itself.

In the light of these observations, it feels unsettling and dreadful to be witnessing the war unfolding in Ukraine and in the Middle East, not just because of the destructivity human beings are capable of but also because, if we look deeper into it, the choice has been that of directing our gaze on the other's death instead of protecting our earth. It looks as if humankind makes the destructive forces prevail over the reparative and vital abilities, as if making life prevail over death were an excessive request. Furthermore, as a result of political cynicism, it is not convenient to make the reparative and vital forces prevail over the destructive ones. Besides, the human being, if affected by a sense of omnipotence, cannot tolerate his/her smallness in relation to the sublime dimension of nature.

Rulers who fuel hatred by inciting groups, collectivities, and sometimes the masses are so caught up in seeing the other who is different from themselves as a possible enemy that they become blind out of indifference, omnipotence, and stupidity, when confronted with the different natural changes they themselves have produced. These changes, despite being perhaps imperceptible and invisible in the beginning, have in recent decades become macroscopic and very frequent; it is as if the effects of climate change and degradation are crying out for revenge for the badly treated environment. If the "narcissistic polarity" (Bion 1992, 105) prevails on the "socialistic polarity" (ibid.), the risk is of not considering that nature has an extraordinary vitality in the powerful energy of the winds, the sea, the melting glaciers, and so on.

Therefore, moving from a mainly theoretical consideration to the real situation and to fieldwork, the discourse becomes tangibly tragic. Since communication has become universal, there is a common world, everybody's experience is simultaneous, and a common death has become possible (Luhmann 1990): globalization is characterized by a ceaseless "space of flows", where different countries act in completely different ways and under the aegis of national egocentrism, where leaders who stir up the primitive masses authorize themselves to the free expression of their greedy and colonizing instincts instead of ensuring the development of the "polis" through the protective function of the common good and a regulatory function which should contain and prevent destructiveness and continuous agitation.

People used to say how much the pandemic would change us in the sense of making us more aware and sympathetic about the environment, even making us kinder and more capable of being in contact with ourselves, of giving more importance to our feelings, to solidarity, to being rather than having. However, if we look at what has been happening in recent times, maybe since the beginning of the lockdown, we realize how various governments are scarcely willing to build a common and – not localistic and nationalistic – sense of belonging, and how much speculation and crime in general affect the health of humankind and of the planet.

References

Bion, W.R. (1961). *Experiences in Groups and Other Papers.* London & New York: Routledge, 1969.

Bion, W.R. (1992). *Cogitations.* London & New York: Routledge.

Corrao, F. (1998). *Ombre*, Vol. II. Milan: Cortina.

Ferro, F.M. & Dionigi, R. (2020). *Non È la Prima Volta. Epidemie e Pandemie. Storie Leggende e Immagini.* Busto Arsizio: Nomos.

Freud, S. (1927). The Future of an Illusion. *SE* 21.

Freud, S. (1930). Civilization and Its Discontents. *SE* 21.

Gaburri, E. (2014). *Navigando l'Inconscio. Scritti Scelti.* Milan: Mimesis.

Kaës, R. (2009). *Les Alliances Inconscientes.* Paris: Dunod.

Kaës, R. (2012). *Le Malêtre.* Paris: Dunod.

Lévy, B.-H. (2020). *Ce Virus Qui Rend Fou.* Paris: Grasset et Fasquelle.

Luhmann, N. (1981). *The Differentiation of Society.* S. Holmes & Ch. Larmore (Trans). New York: Columbia University Press, 1982.

Maffettone, S. (2020). *Il Quarto Shock. Come un Virus Ha Cambiato il Mondo.* Rome: Luiss University Press.

Malidelis, D. (2019). "Interview on Psychoanalysis and Ecological Crisis with Cosimo Schinaia". Spiweb.

Mann, T. (1924). *The Magic Mountain.* London: Penguin, 1969.

Neri, C. (1995.) *Gruppo.* Rome: Borla.

Ricoeur, P. (2000). *Memory, History, Forgetting.* K. Blamey & D. Pellauer (Trans). Chicago, IL: University of Chicago Press, 2004.

Sontag, S. (2003). *Regarding the Pain of Others.* New York: Farrar, Straus and Giroux.

Chapter 6

Catastrophe versus Catastrophic Change. Between Psychoanalysis and Art

Luca Caldironi

Introduction

Among the physiological difficulties of human beings, there is one that relates to the possibility of seeing and observing events that affect us when they happen. How, then, to deal with the Zeitgeist in which we are immersed? It is important to initiate a work that involves the possibility of reflecting on the experience we are living and, in so doing, favoring a perspective dimension of it.

When we speak of psychic reality, we mean a reality that resides within the subject, and which conditions what he has doing in the world with desires and fantasies. The environment that surrounds us does not invite us to listen to our inner world, and our attention is increasingly focused on new media rather than on deepening the emotional contents and their meaning.

Freud's quote at the beginning of his essay "The Interpretation of Dreams" (1900), *Flectere si nequeo superos, Acheronta movebo* remains a reference point for talking about the relations between artistic creativity and psychoanalytic research, especially Bionian.

In a world where the term "catastrophe" is increasingly used as a sign of an impending apocalypse, it is important to recover its etymological meaning. This term, in ancient Greek, is composed of *kata* (down) and *strophein* (turn, turn around), indicating a kind of reversal.

In the structure of Greek tragedy, the part indicated by "catastrophe" is associated with a revelation with respect to an unknown fact, a revelation that may be followed by a "catharsis." A term, the latter, which in psychoanalytic culture (Freud and Breuer 1895), is associated with the liberation of incrusted and conflicting emotions and the putting back into circulation of "free psychic energy" and thus re-investable. This digression allows to juxtapose the threat of a universal catastrophe with the concept of "catastrophic change" proposed by Bion.

With this concept Bion gives to catastrophe also a positive meaning. According to him, it is a phenomenon that marks an evolutionary leap, both in terms of individual mental growth as well as in terms of group dynamics

DOI: 10.4324/9781003498605-6

and social transformations. Francesco Corrao (1974) points out how Bion had noted that the "lie" can be used as a defense against the risk of a psychic change perceived as an unexpected and sudden emotional upheaval, a catastrophe, in fact. But, we add, perhaps only a catastrophic change can save us from catastrophe!

In his analytic work, Bion has shifted the focus on the fact that the interpretation should not only be based on prior knowledge but should emerge from the analytical experience itself. This new point of view could be considered a significant change in traditional analytic practice. This change could be seen as catastrophic in the sense of a break with the past and a way of placing more emphasis on emotional experience and the value of emotions as vehicles for understanding unconscious thoughts.

The ability to express emotions, especially mental pain, is a fundamental ontological concern. The degree of capacity to tolerate or suffer pain varies from person to person and that it changes over time and from one culture to another.

The word "suffer" comes from the Latin *suf-ferre* (to put or take something in), which makes "suffering" different from "feeling", which is closer to the realm of the senses and, like touch, closer to the receptive surface. The term "suffering", with its implied internal space, evokes a spatial dimension, as the experience of suffering moves from the surface to an internal space in which all the forces at play converge. In Italy, and particularly in Venice, where I currently live most of the year, we could say that they meet in the *campo*. In Venice, *campo* means a public area that can be of any size or shape. Throughout the city's history, these open spaces have offered the inhabitants opportunities to develop relationships and exchanges. Unlike the Greek agora and its centrality in the life of the polis, our multiple *campi* evoke the multiple sets of the classical Venetian *commedia dell'arte* [1], which was staged throughout the city, rather than the single stage of ancient classical tragedy. By analogy, these fields, these spaces within, between and among us, are available to the human fabric to which they belong. With a whimsical thought I like to imagine these "fields" as emotional geographies, anticipating what would later become "field theory" from Bion onwards. They are spaces of sharing in which stories can arise that give shape to heightened emotions and feelings that can lead to catharsis, in a continuum of concealments and revelations. As Bion observed about dreams, the core is not the manifest content, but the emotional experience. The same applies to narration.

Michael Eigen (1998) reminds us that we communicate through narratives. We impose an appearance of coherence and integration. But on what do we impose this coherence and integration? What are the broad emotional experiences that we filter through narratives, that we create through narratives?

We try to give ourselves messages from deep within. And we try to do this through narratives that distort them and, in a sense, hide them. Freud was also interested in the way the dream hides emotional messages. Bion believes that we cannot escape this phenomenon. With the imposition of coherence and narrative integration, something is lost, perhaps distorted, but also something is created to work with.

Pieces of emotional experience can be fed into unconscious waking thought and common sense. We can build – good or bad – with the emotional sense distilled from dreams. Art, myth, politics reflect the ways in which emotions are organized, recreated. But it is good to realize that something unknown about emotional reality remains.

Interestingly, *narrare* is a verb from the ancient Greco-Roman *gnarigare*, which is a compound derived from the Sanskrit words *gna*, meaning "to know", and *gare* (Latin *agere*), "to do or act", in the sense of "to make." Narrating is thus "making aware", which implies a relationship of narrating and listening that restores to its separate constituents – the narrator and the listener – the possibility of self-knowledge and reciprocity. Along these lines, we can imagine a language of reciprocity, a discourse that comes from both sides, in chronicity as in emergency. In the context of a narrative medicine practice, such a language requires, both caregivers and patients, as well as family members, to engage in a process of speaking and listening that makes space for the disorganized "roar" of emotions, particularly painful ones, while moving towards a possible shared coherence. Facilitating this level of communication, which considers the emotional listening of both caregivers and patients, requires a more complex type of training, which through a more careful modulation of therapeutic distance, allows the experience of suffering to give rise to the shared and separate narratives of both patients and caregivers.

Faced with these situations, defense mechanisms such as avoidance and detachment are an almost inevitable response. These phenomena, together with others such as manic or hypomanic defenses, make difficult the contact with the internal reality. Contact which, on the other hand, is a prerequisite for working through the changes taking place and tolerating frustration and feelings of helplessness. Nobody is exempt from these difficulties, and what we can try to do as analysts is to keep alive as much as possible a potential and/or transitional space within which the possibility of thinking remains alive.

The Unconscious in Social Life, Several Questions

David Morgan (2019) writes that in "Civilization and Its Discontents" (1930), Freud was pessimistic about man's cruelty to himself and the tendency to eliminate anxieties about life and death through war and power. The same happens with regard to man's blindness to what we are experiencing, both climatically and from a socio-humanitarian point of view. Consequently, it becomes important to start from the microcosm represented by the real and metaphorical analytical room and extend our gaze/experience beyond this. Only by being able to better understand and work through the conflicts in our inner world, will we be able to act more responsibly in the world around us and in our coexistence with others, preventing the group and/or the unconscious crowd from taking over the individual's capacity for discernment.

While it is clear that it is important to be able to intervene on these forces that run through society, we know how difficult this operation is. It is a question of approaching forces that interface with our psychic defense mechanisms that are deputed to the removal of all that is undesirable and that creates unease.

But it is equally true that this need to split off comes to interfere with the possibility of facing reality as it is.

Pursuing the awareness is an operation that involves frustration and the ability to tolerate moments of pain and bewilderment. In this regard, in "Learning from Experience" (1962), Bion tells us how precarious is the human being's capacity to tolerate the truth about himself and how, consequently, the mind is always ready to create lies to oppose truth.

In "Cogitations" (1992), Bion discusses the importance of mental health in facing reality and achieving personal development. He emphasizes the necessity of constant fact-finding and avoiding distractions to maintain mental well-being. Bion points to the suggestive comparison in which, just as the Earth is protected from the possible harmful effects of cosmic rays, so too does the human being need a protective atmosphere against harmful influences.

In "The Question of Lay Analysis" (1926, 248), Freud tells us:

"As a 'depth-psychology', a theory of the mental unconscious, it can become indispensable to all the sciences which are concerned with the evolution of human civilization and its major institutions such as art, religion and the social order. It has already, in my opinion, afforded these sciences considerable help in solving their problems. But these are only small contributions compared with what might be achieved if historians of civilization, psychologists of religion, philologists and so on would agree themselves to handle the new instrument of research which is at their service. The use of analysis for the treatment of the neuroses is only one of its applications; the future will perhaps show that it is not the most important one. In any case it would be wrong to sacrifice all the other applications to this single one, just because it touches on the circle of medical interests."

Extending Freud's remarks, we come to hypothesize the possibility of thinking of a new meta-psychology. A meta-psychology that, as was the case at the beginning of the discoveries about the value of the unconscious and how decisive it was for the understanding of the human being, can now, through the extension of psychoanalytic practice to multi-psychic devices call into question the conception of another model of intelligibility to account for the plurality of places, dynamics and economies of unconscious psychic reality that emerge in such devices (Kaës 2015).

From Bion's early experiences in his observations of groups, one begins, in fact, to speak already of the group mind and the plurality of each individual. The Copernican revolution already initiated by Freud regarding the loss of centrality of the individual is extending. It was another "catastrophic change"!

It is appropriate, therefore, to remain faithful to Freud's thought by declining it on the variations of experience that are produced in the process of progressive knowledge of the unconscious. Knowledge that, as can be guessed, can never be completed (Caldironi 2018).

The internal images (sometimes described as *imagos*), gradually accumulate, and it is this accumulation that creates the conscious mind. These hypotheses or explanations of what happens to us can only be derived from what we already know. In the light of further experience, we adapt these assumptions; in other words, we change the internal images. Sometimes, however, these assumptions do not change in the light of experience. Instead, they become fixed as unconscious beliefs and appear as facts in the conscious mind.

When Clare Britton was little, an older boy told her that Father Christmas did not exist. She immediately realized that he had assumed that Father Christmas was a fact and not a belief. She wondered what other things she had assumed to be a fact could turn out to be just a belief. Since unconscious beliefs are experienced as certainties, they protect us from the need to think and, in so doing, reduce anxiety (Britton 1950).

This mechanism reduces our capacity for discrimination and self-orientation, favoring a vision that necessarily remains more superficial.

This superficiality has resulted in a progressive procurement by increasingly powerful technological and communication systems over the human element, which appears increasingly at risk of dis- or misinformation in an era when the power of Information reigns.

Christopher Bollas (2018) stresses the difference between sight and insight. According to him, some observers believe that this is the age of the "display": we seem to be attracted by the images of life in the universe mediated by technologies. And although we are informed, visually, and possess memories of what we have seen, our insight is relatively poor. For there to be insight, it is necessary for consciousness to be directed towards the inner world and for there to be interest in the various meanings of our lived experience. Bollas calls the use of sight to avoid insight *sightophilia*. A person who tends to look rather than think is a *sightophiliac*. Refractive thinking, therefore, eliminates meaning.

Mauro Ceruti (2014) reminds us that today's great global crises are at once social, health, political, economic, and concern both the relationship of populations to each other and the relationship of humans to the Earth. Complexity is a word that has been misused lately. But let us remember that it derives from the Latin word "complexus," which means tight, woven together: it means that the various dimensions cannot be separated. We have to think of complexity as in a work by Escher and follow Calvino's lesson on the task of representing the world as a tangle, without attenuating its inextricable complexity.

Italo Calvino (1988, 84) goes on asking with himself "what will be the future of the individual imagination in what is called "the civilization of image"? Will the power of evoking images of things that are not there continue to develop in a human race increasingly inundated by a flood of prefabricated images? [...] Memory is littered with pieces of images, like a garbage dump, and it is increasingly unlikely that one form among many can be distinguished."

At the same time, we know how important the value of the image, imagination and the physical and emotional 'mirroring' it entails is. Back in the day, Hume told us that:

"In general, we may observe that men's minds are mirrors to each other, not only because they reflect each other's emotions, but also because those rays of passions, feelings, and opinions may be often reverberated, and may decay by insensible degrees."

(Hume 1739, 2.2.5.21)

This thought emphasizes the importance of having multiple points of reference to avoid collapse.

Contemporary science too shows that it is possible to make significant strides in understanding human being through an interdisciplinary approach. Hence the importance of establishing a dialogue that gives space, time and place to this process and the importance of having and/or creating a container of this process.

We now want to dwell on creativity and its necessity through a psychoanalytic perspective. The term perspective indicates the use of a point of view that makes use of a particular knowledge. We also intend to focus on a concept that does not want to be limited or confined to one form of creation, but to capture that place where the emotional encounter is articulated between several phenomena at once. These are phenomena that intertwine our emotional and neuro-sensory worlds and can be promoted by a fruitful encounter.

Holger Kalweit (1988) stresses how important it is to keep in mind that healing means daring to step outside one's own fence, but this stepping outside is far from easy and concerns all fields of knowledge. This is why it is important for each field of knowledge to ally itself with the other, not out of a form of diffuse eclecticism, but out of an openness to the world; an openness that comes to involve and intertwine concepts such as approximation, intuition and intention, defined or declined within the creative process.

For these different framings to become probes suitable for investigating new territories, it is necessary that the field of observation also widens and contemplates an asymptotic approach to a project that can only proceed through transitory and provisional instances. It is an exploration "of the adjacent" that can prepare successive stages.

Bion (1983) tells us the importance to discover processes by which to forget what has been learned, in order to be able to be sensitive to the fundamental thoughts and feelings that may have survived – there may still be some traces of wisdom in the human race.

There is indeed a difference between promoting an increase in defenses ("help to be stronger," "get out of the difficult situation!"), and the possibility of "letting go," despite the terror, in order to be able to "enter the situation" more and more and listen to oneself and one's fears.

This is the conflict to be kept alive and we know well how bitter and difficult it is and how seduced by oblivion.

In this regard we explore the significance of desire in society, distinguishing it from mere need. We suggest that desire, as a human tension, drives us to engage with the world and confront the unfamiliar.

Between Utopia and *Eutopia*: Give a Chance to Desire

Let us now consider, in addition to desire, other feelings that are fundamental to restoring meaning to our time, to our "temporality," to give an "emergency" to our crises, such as enthusiasm and passion. Winnicott (1971) said, referring to himself: "Oh Lord, may I be alive when I die."

We might add: "may I be and remain alive as much as possible in my life." For this to be possible, two other ingredients, such as enthusiasm and passion, are essential. These are not easy feelings, but opportunities to be cultivated with care and perseverance.

"Enthusiasm," as the word itself (*en-theos-ousia* / with-God-in-itself), informs us, is a state of mind in which one experiences a passionate feeling. It can be a passionate feeling for some ideal or even a kind of heated excitement towards life itself that expresses as a continuous push in research or experimentation. This feeling is profoundly distant from any form of fanaticism or excitation by the use of psychotropic drugs. It represents a natural gift, although it is obviously related to the environment and the type of experiences that occur within the environment.

They are feelings that imply a relationship with a sense of belonging-alterity that generates tension. In enthusiasm there is a tension between the subject and an internal "life-giving" object. In desire, a kind of "aspiration towards." In the passion, a deep feeling that ethically "co-inverts" subject-object into what we can call, with Martin Buber's expression, an "I-Thou."

Bion's (1970) "act of faith" is the container that encompasses all the feelings we have mentioned. Faith possesses a propulsive force that can help the analyst to support his desperate or depressed patient, but it can also help every human being in those moments when desire is lacking, and it is necessary to place oneself in confident expectation.

To face this experience, the human being must necessarily confront his or her own sense of loneliness, that sense of loneliness that accompanies us from birth, but which must not turn into anguish. Anguish is difficult to share, it is a "place" where time is "stifled." That is why it becomes important that anguish is transformed into pain, into mental pain. Pain involves the sense of time, "rhythm", not only in relation to its working through, but also as a defense of a "good object." For this reason, it becomes important to have a place where to express it, to find that emotional container that makes thinkable and nameable that feeling of dismay that Bion called "Nameless Terror."

These considerations seem to us profoundly relevant and useful in a civilization, such as the one we live in, which tends to deny death and its mystery. Even more, they seem useful to us in order not to risk disregarding, thus denying, the problem of the possible progressive destruction of our planet, of its vulnerability which, like ours, can be denied. Hence the importance of not losing the value of hope; Bion (1970) uses the term "patience" to indicate that feeling that allows us to pause in an expectation fueled by hope. An expectation that is co-equal with a certain amount of suffering, but also tolerance with respect to frustration. The word "hope", also in its etymological root, which in addition to Latin *spes* also has the Sanskrit root *spa* in which we find the meaning of "tending towards a goal," offers us the image of a "tension," of an "active suspension."

This distinction is also emphasized by Isaac Tylim (2019) when, distinguishing between abstract and concrete hope, he says: "In contrast to abstract hope, concrete hope is closer to the present in its participation and awareness of present obstacles. While abstract hope can be seen as a manic defense, concrete hope embraces the limitations experienced in the present."

Desires are informed by the real; this means that it is concrete in the sense that this form of hope recognizes what one feels is lost or damaged, intimating the depressive position. Concrete hope aims at integration and repair.

It is about moving forward despite disappointments, promoting activism by challenging passivity and detachment.

However, it is important to note that the relationship between disaster and hope is not automatic or guaranteed. History shows us that human beings have a remarkable capacity for adaptation and resilience and often find hope even in the most difficult situations, but this capacity for adaptation also needs to be supplemented with the ability to work out new parameters and make room for new possibilities in thinking.

By a sort of transitive property, the difference we have just highlighted is also expressed in the different value we can give to the term "utopia," Tylim again reminds us that utopian imagery can offer a lifeboat for navigating turbulent waters. Utopia helps to survive the impossible present, while charting the course for a new and different future through hope of a concrete kind.

Hope as fuel for utopia: hope is often seen as a motivating force that fuels the search for a better future and an ideal image of a perfect society; without hope, utopia may seem unattainable and not worth striving for. On the other hand, utopia can sometimes lead to disappointment. When expectations for a "perfect future" are too high or unrealistic, reality can disappoint people. This can lead to a loss of hope. This is not an easy balance because the dynamic forces are intertwined.

And so, while we are aware of the meaning of "utopia" in its meaning of an unrealistic project, we favor that of "representative of an absence." Etymologically "utopia" means "non-place" or rather, a non-place that becomes a pole of attraction. It is a form of use of the "negative" which, as in Bion's (1967) considerations on the difference between "Nothing" and the "No-Thing," makes a sense of presence implicit. This "non-place" comes to represent a path in which desire tends to push the boundaries. A "utopia" that, playing with the homophonic "eu-topia" of the English language, can tend towards a "benevolent place," a "eu-topos" in fact, and not only in concept. It therefore becomes important to rediscover the meaning of utopia, a utopia understood as a space for creativity and life, and to recover a vision of the future that becomes a vehicle that spans time, tracing dimensions that have ontological potential and involving values such as awareness and a greater sense of responsibility. Utopia becomes a precious refuge in the expectation and participatory search for a possible future.

From a psychoanalytic point of view, we can attempt a translation of this concept through the differentiation between "Ego-Ideal" and "Ideal-Ego." Starting from an "original narcissism," two different paths can be distinguished. The one represented by the "Ideal-Ego" remains a continuation of the feeling of perception of wholeness that characterizes original narcissism. A large share of idealization of the self is maintained. In the other, which we call "Ego-Ideal," there is a shift of this idealized perfection outwards, first towards parental figures and later extended to the interrelated environment.

The shift, therefore, from the Ideal-Ego to the Ideal of the Ego comes to take place within an investment that is not only more mature, but also more realistic towards relations with the environment. The perception of a loss is already implicit in this passage. A narcissistic loss given by the investment in the external world. The drive to process this loss becomes the indispensable push to activate the tension towards "something", towards a "project."

These concepts will have a further development through Bion's (1961) research and his work on groups. He will come to hypothesize a polarization between being in a group (man as "political animal" according to Aristotle), and the individual.

With the terms "narcissism" and "socialism," Bion describes the forces that push inwards (ego-centric), and the forces that push outwards (socio-centric), warning against excesses on one side or the other. Only a sufficiently elastic ego can be willing to manage this physiological conflict and thus give "home" to a world of emotions and its different currents.

When we speak of psychic space, we do not mean a single dimension, but multiple areas of reality that, proceeding from an intrapsychic dimension, pass into areas that concern the space between the subject and his or her environment, the couple and/or the family group, up to a more extended space that is intertwined with social and cultural space.

This topic delves into the relationship between neuroscience, art, and creativity. Antonio Damasio's (1994) concept of the "body-as-if loop" suggests that our brains mimic the actions and emotions we observe in others. This idea is applied to neuroaesthetics, proposing that our mirror neurons activate when viewing art, enabling us to understand and emotionally respond to it.

David Freedberg (1989) emphasizes the importance of studying public reactions to art rather than defining creativity itself. The concept of creativity is further explored through Max Ernst and Italo Calvino, pondering the potential of art to transcend individual perspectives and give voice to the non-human. This last term contemplates the evolution of human identity in the face of technological advancement and the blurring boundaries between self and other.

It is a boundary line, a threshold (a "caesura" Bion would say) that is increasingly labile, transitive and transient. On this line and on this 'caesura' the dice will be rolled for the future, but beware because experience teaches us that already *alea iacta est* (the die is cast)!

The Importance of Keeping Unsaturated Space

Proceeding along this line has the ambition of bringing back into play aspects that favor tapping into potential that, an excess of "determinism" has limited over time. This is not a simple wish, but it is a wish that requires the commitment of a collaboration between different specificities and a change in the very mode of learning. In this regard, Luigi Boccanegra (2018), quoting Yves Bonnefoy, argues that the need for clarity and control must not saturate the child's initial disposition to potentially learn several languages, before a codified mother tongue fixes for him a definite correspondence between the sounds and concepts of his culture.

It is a cultural "decolonization" that opens up the possibility of proceeding by images, as in the dream, and offers space to an "over-thinking" that, with its going "against the tide," makes peace with its debt to non-univocal knowledge.

Within this complexity, it is the system itself that, through a moment of "breakdown," can offer us the opportunity to use a different point of view. Mauro Ceruti (2014) tells us that two eventualities can occur. In the first, the engine of development is the "knowing of not knowing." In this case, new discoveries, new contents are placed within a mental space that remains firm and invariant. The world remains the same and it is only the knowledge of its reasons that extends, expands, deepens. Staying at "home," one goes and sees further and further, by means [however] of relatively linear extrapolation procedures. The second type of development, which is rarer [and, we might add, probably more valuable], is based precisely on the often sudden or even dramatic experience of "not knowing that one does not know." It is an expression that reminds us of Freud's when he says that the ego is not master in its own home. This experience calls into question the very established mental space in which one expected to be able to place the discoveries and contents to be acquired in a non-problematic way. Such an experience, in fact, imposes the process of "learning to learn," demands a change in the modes of learning, the transformation of the very types of questions.

A radical, indeed, catastrophic change is needed! Lorena Preta (1990) says that change is experienced as catastrophic before being experienced as growth and conquest.

Ceruti (2014) again tells us that our species has never possessed a preventive wisdom with respect to the consequences of its innovations: rather, it has had to learn, at its own expense, the new limits and possibilities. Ceruti proposes, on the possible origins of the repetition compulsion, the difficulty to prevent disasters and hypothesizes that mankind today could therefore be the victim of the secondary effect of an evolutionary process triggered by environmental conditions that are no longer current. (Conditions) that would have enhanced the mechanisms of internal coherence between groups, to the detriment of the mechanisms of interaction between groups and with the environment in general.

Modern-day humanity would be the victim of a kind of legacy of the past. Perhaps a first conscious humanity is on the verge of being born of having a decisive responsibility to nature and to having to think together the one and the many, identity and diversity, and this for its own survival.

Our society has, therefore, the need for a real "re-humanization." It needs to re-think itself, no longer through the interminable contentions of small groups, but through the multiplication of connections that lead from the single individual to a single planetary totality, through multiple and disparate collectivities.

Bion too warns us that the human being's ability to tolerate "truth" about himself is always precarious. Therefore, the mind is always ready to "mentire" (to lie) in order to remove this pain from itself. The pain of "thinking." Hope and trust are the other elements that flank the possibility of thinking and offer physiological help in dealing with frustration and the experience of pain. If thinking can be considered the most evolved form of the self-pre-servation instinct, we must, at the same time, ask ourselves how it is possible that this capacity has evolved and is currently evolving into such lethal activities for human beings. Paraphrasing Bion, one could say that intelligence alone, without wisdom, is not sufficient to curb blind destructiveness. It is an aggressive destructiveness that proceeds heedless of the Other, Other also understood as Nature and Environment. Our personal world, Stefania Nicasi (2023) reminds us, is interwoven with otherness, language, memory, history: there is no such thing as a degree zero. This is true for the individual as well as for communities and generations. And so, it is that:

> "The priority seems to become not the advent of the human justice of universal well-being, but the prevention of a planetary catastrophe, nat-ural or vital, that could put an end not only to history but to humanity itself. The theme of the 'end of History' no longer takes on, in this beginning of the new millennium, an eschatological sense, nor even … that of fulfilment (Hegel), which it still retained at the end of the last century, but a catastrophic and deadly sense. […] Death, driven by the icy wind of the 'Damoclean threat' of great catastrophes and the night-mare of an apocalypse of human life on Earth, is once again gaining ground within our own psyche."
>
> (Ceruti and Bellusci 2023, 383–384)

Art like a Psychoanalytic Compass

Let's try to find help using art, creativity and psychoanalysis as a compass/intuition for the future.

> "In an age where meaningful global communication and intelligent restructuring of our environment are imperative, art can play an impor-tant role. It can influence the collaboration and integration of disciplines, and can offer a useful approach to problems. A well-designed work can motivate people and influence their perception of things."
>
> (Denes 2019, website)

By creative activity I mean any human activity that by various ways pro-poses an "art-ifex" (creator) of something new. Something that transits in that space that exists between desire, will and the possible realization. The creative process is thus an opening of human being towards his future, enabling him to change his present.

Bion (1967) says that he fully understands that there can be a crisis of development in which the human being is absolutely terrified that the future is unknown, cannot be known by him in the present moment, and can only be known by certain people, described in terms of "genius," or "mystic," who have a special relationship with reality.

Therefore, it is appropriate to combine the use of the psychoanalytic method with art and more generally with "creativity" as an attempt to approach and reveal deeper aspects of the human being.

The American artist Agnes Denes reminds us: "We live in an age of complexity, in which knowledge and ideas present themselves faster than they can be assimilated, while disciplines progressively move away from each other through specialization. Laboriously acquired knowledge accumulates undigested, blocking meaningful communication. A clearly defined direction for humanity is lacking. [...] Making art today is synonymous with taking responsibility for our fellow human beings."

In both psychoanalytic and artistic work, intuition is a unique and fundamental moment. It is necessary, therefore, to concern ourselves with how this moment can be fostered and how we can free our intuitive capacity from all those co-presences that burden it. It is a far from easy task and the very concept of 'intuition' presents itself to our eyes as a 'chimera' that can be approached and never grasped!

I purposely used the term "chimera," having in mind Dino Campana's beautiful poem "Chimera" in "Orphic Songs" (1914). Chimera, the mythical figure meaning a hybrid animal, is the dream chased and yet unattained, evoking poignant nostalgia, contemplation of mystery.

Still regarding the concept of intuition, it is interesting to recall how in the exhibition "Intuition," held in Venice (2017), curator Alex Vervoordt, speaking about the theme of the exhibition, says: "[Intuition] is a sensation that comes from total freedom, from being at one with cosmic energy. It is knowledge before knowledge. It is understanding before understanding. Intuition is the result of processing a large number of messages that arise in our brain and body, and with attention we suddenly see the light. Intuition gives us new ideas and does not always tell us where they come from."

It is no coincidence that these both considerations come from the world of art and are expressed through metaphors. They represent, in fact, the desire to express the ineffable and at the same time recognize its important value. This is what artistic expression and analytical work have in common, the possibility of experiencing and at the same time being able to communicate an emotional experience.

In particular, reading Bion's work as a watermark, I agree with the statement by Rudy Vermote (2011), quoted by Adela Abella (2016, 453), when he says: "It has been said how Bion increasingly evolved, in his late work, towards an unexpected and suggestive aesthetic conception of psychoanalytic work, one that explores 'the profound and formless infinite' of the unrepresented, undifferentiated level in search of ineffable truths and that must draw on intuition and faith. [...], Bion, speaking of his last work: 'I was forced to seek asylum in fiction'. Disguised as fiction, truth occasionally slipped out."

This last sentence, indirectly, comes to renew how important it is to use different expressive means in order to be able to "leak out" glimmers of "truth."

These are operations that do not help in the avoidance of frustration, but rather exploit frustration as an engine to seek new expressive possibilities involving different forms of experience and emotions. Both internal and external interlocutive elements are favored. By internal, we mean listening to one's own personal experiences, such as those of the artist with his individual history, his historical context, and by external, we mean listening to the response of the person who enjoys the work and the environment. Following this thought, we can see how much what is proposed when we talk about contemporary art comes close to Bion's concepts. In particular when he speaks of the need for the analyst to maintain, as much as possible, an "unsaturated attitude", "without memory and without desire." An attitude that leaves room for new ideas, associations and possibilities.

An example that can help us to proceed in this discourse is the comparison that Abella (2012) makes starting from Bion, as analyst and John Cage, as artist. It starts from a very "pro-(e)vocative" statement, as Bion would say, by Cage: "The function of art is to conceal beauty: this has to do with opening our minds, because the notion of beauty is only that which we accept." The final consequence of this line of thought is Cage's suggestion of an attitude of renunciation of "one's own desire, pleasure and memory." It should be noted that these parallels between Cage and Bion are not casual and superficial but are rooted in their deep values concerning art and psychoanalysis, respectively. Cage's renunciation of desire, beauty, and memory, as well as Bion's warning about desire, memory, understanding, or coherence are nothing but the basic conditions necessary to open oneself to new truths, both in the external world and in the mind. In this in-certainty search, contemporary art meets Bion's proposal for a position of openness and tolerance, one that allows one to pay attention to "imaginative conjecture," "speculative imagination," wandering thought, "wild thoughts," "interference," and "unborn children" (Abella 2012).

These "invitations" are indispensable elements for dealing not only with the complexity of our time, but that they can also stimulate and broaden our perception of the world, in particular the complexity of the world in which we find ourselves living. We know, in fact, how difficult it is to have a panoramic vision of time that brings us up to date and how desirable it is to take "strategic diversions" and have multiple points of view.

Bion had borrowed from the poet John Keats, of "Negative Capability", meaning that capacity to "pause" in experience and not precipitate hasty and defensive explanatory short-circuits. But we also find it useful to associate with this concept those aspects of contemporary art which, as in Robert Rauschenberg's White Paintings and later in Cage's famous piece 4'33", have the ambition of "disappearing" to leave space for a "whole" that forces the observer or participant in the event to always seek something Other, to move from an apparent simplicity, without color, without notes [...] towards deeper complexities that are unreachable because they are always in progress. Another similar example can be the use of "negation" within both the iconographic process and Marcel Duchamp's use of form.

Adela Abella (2007, 1048) reminds us that "Duchamp engages in what one critic has called 'reaching the limit of the unaesthetic, the useless and the unjustifiable': he works on a geometry of the fourth dimension in search of the 'infra-subtle' and the timeless. He imagines the dictionary of an unknown language whose alphabet would be composed of film endings, sells 'shares' in a gambling system destined to beat Monte Carlo roulette."

Cage takes a different route. While expressing his distrust of language in terms very similar to Bion's, he turns not to science but to chance – an insidious manoeuvre already used by Duchamp. Thus, he simply composes following the Chinese divination book *I Ching* or plays on a 'prepared piano' in which all sorts of things are added to the strings. The goal is always the same: to produce an unpredictable series of sounds, which he sees as a radical way of freeing himself from beauty, memory, and desire (Abella, 2012).

An example celebrating a common ground between Duchamp and Cage is the multi-sensory performance "Reunion." A performance in which, through a chess game between the two, an event is staged that is musical, acoustic and environmental at the same time, a true aesthetic of indeterminacy. A game "without purpose" other than that of a con-versation between all participants, subjects and objects included. It is a form of communication that increasingly distances itself from the rational knowledge model and becomes more and more extreme. It is reminiscent of Bion's later writings, in which language becomes increasingly "pro-e-vocative" and generative of conflicting feelings. It is a language, both that of the artists mentioned and that of Bion, that is "catastrophic", a language that expresses a profound rupture of "hinges."

The convergences we are highlighting represent the value of the dialectic between different perspectives and can be imagined as the "content" always in progress within a "container" that allows its evolution. It is an interdisciplinary dialogue that tolerates the succession of moments of illusion and disillusionment, of respect for tradition and daring new creative paths or, as we like to say, "cre-active"! With this word we want to emphasize the "active" aspect necessary for a deeper exploration of the self, whether this concerns the psychoanalytic or artistic area.

It is often not sufficiently considered that creativity requires a certain amount of aggression. Energy needed for a process that requires constant gravitating between "chaos," "destruction," and "reworking" to forge the inner and outer reality. This is the possibility of a change that Bion describes as "catastrophic," but which can open up new dimensions of experience.

This process can be linked to what happens in the analytical path when we work in the transference and counter-transference dimension going beyond the purely hermeneutic aspects. To be in conversation is to be beyond oneself, to think with the other and to return to oneself as other.

It is a "conversation" in "conflict", insofar as it re-experiences the tension between fear of the "unknown" and thus the defensive need to remain in the area of the already acquired, and the desire to reach out and get to know the most authentic parts of oneself and to open oneself up to the new and the unknown. It is precisely for this reason that it is useful to join forces between art, and in particular contemporary art and psychoanalysis, to give the impetus to our navigation by sight!

Only the attainment of "mental growth can foster new possibilities to think and create in an authentic personal way.

Note

1 *Commedia dell'arte* is a theatrical form characterized by improvised dialogue and a cast of colorful stock characters that emerged in northern Italy in the 15th century and rapidly gained popularity throughout Europe. In its golden age, plays of the *commedia dell'arte* (literally, "comedy of professional artists") were usually performed in open air by itinerant troupes of players. Performances were based on a set schema, or scenario – a basic plot, often a familiar story, upon which the actors improvised their dialogue. Thus, actors were at liberty to tailor a performance to their audience, allowing for sly commentary on current politics and bawdy humor that would otherwise be censored.

References

Abella, A. (2007). "Marcel Duchamp: On the fruitful use of narcissism and destructiveness in contemporary art". *International Journal of Psychoanalysis*, 88(4): 1039–1060.

Abella, A. (2012). "John Cage and W.R. Bion: An exercise in interdisciplinary dialogue". *International Journal of Psychoanalysis*, 93(3): 473–487.

Abella, A. (2016). "Using art for understanding of psychoanalysis and using Bion for the understanding of contemporary art". In H.B. Levine & G. Civitarese (Eds), *The W.R. Bion Tradition* (pp. 451–466). London & New York: Routledge.

Bion, W.R. (1961). *Experiences in Groups*. London & New York: Routledge, 1969.

Bion, W.R. (1962). *Learning from Experience*. London & New York: Routledge, 2023.

Bion, W.R. (1965). *Transformations: Change from Learning to Growth*. London & New York: Routledge, 1984.

Bion, W.R. (1967). *Second Thoughts*. London & New York: Routledge, 1984.

Bion, W.R. (1970). *Attention and Interpretation*. London & New York: Routledge, 1984.

Bion, W.R. (1979). "The Dawn of Oblivion". In *A Memoir of the Future*. London & New York: Routledge, 1991.

Bion, W.R. (1983). *Italian Seminars*. London & New York: Routledge, 2005.

Bion, W.R. (1992). *Cogitations*. London and New York: Routledge.

Boccanegra, L. (2018). "Il contributo di Yves Bonnefoy alla clinica". *Gli Argonauti*, 156: 13–28.

Bollas, C. (2018). *Meaning and Melancholia: Life in the Age of Bewilderment*. London & New York: Routledge.

Britton, C. (1950). "Child care". In C. Morris (Ed), *Social Case-Work in Great Britain* (pp. 168–181). London: Faber & Faber.

Caldironi, L. (2018). "Psychoanalysis and cyberspace: Shifting frames and floating bodies". In I. Tylim & A. Harris (Eds), *Reconsidering the Moveable Frame in Psychoanalysis* (pp. 233–246). London & New York: Routledge.

Calvino, I. (1988). *Six Memos for the Next Millennium*. P. Creagh (Trans). Cambridge, MA: Harvard University Press, 1988.

Campana, D. (1914). "Chimera". In *Orphic Songs*. I.L. Salomon (Trans). San Francisco, CA: City Lights Books, 1998.

Ceruti, M. (2014). *La Fine dell'Onniscienza*. Rome: Edizioni Studium.

Ceruti, M. & Bellusci, F. (2023). "Oltre il catastrofismo". *Psiche, Finimondo* 2: 381–392.

Corrao, F. (1974). "Presentazione". In W.R. Bion, *Il Cambiamento Catastrofico* (pp. 7–11). A. Baruzzi (It. Trans). Turin: Loescher.

Damasio, A. (1994). *Descartes' Error: Emotion, Reason, and the Human Brain*. New York: Putnam.

Denes, A. (201). *Catalogue: Absolutes and Intermediates*. E. Enderby (Ed.). New York: The Shed.

Eigen, M. (1998). *The Psychoanalytical Mystic*. New York: Free Association Books.

Freedberg, D. (1989). *The Power of Image: Studies in the History and Theory of Response*. Chicago, IL: University of Chicago Press.

Freud, S. (1900). The Interpretation of Dreams. *SE* 4.

Freud, S. (1926). The Question of Lay Analysis. *SE* 20.

Freud, S. (1930). Civilization and Its Discontents. *SE* 21.

Freud, S. & Breuer, J. (1895). Studies on Hysteria. *SE* 2.

Hume, D, (1739). *A Treatise of Human Nature*. London: Penguin, 1986.

Kaës, R. (2015). *L'Extension de la Psychanalyse. Pour une Métapsychologie de Troisième Type*. Paris: Dunod.

Kalweit, H. (1988). *Dreamtime and Inner Space: The World of Shaman*. Boston, MA: Shambala.

Morgan, D. 2019. *The Unconscious in Social and Political Life*. Bicester: Phoenix Publishing House.

Nicasi, S. (2023). "Editoriale: Eredità in Finimondo". *Psiche, Finimondo*, 2: 335–349.

Preta, L. (1990). "Prefazione". In M. Ceruti & L. Preta (Eds), *Che Cos'è la Conoscenza* (pp. VII–XIX). Rome-Bari: Laterza.

Tylim, I. (2019). "Hope, despair, and utopia". *Room: A Sketchbook for Analytic Action 10.19*. https://analytic-room.com/essays/hope-despair-utopia-isaac-tylim.

Vervoordt, A. & Ferretti, D. (Eds) (2017). *Intuition*. Catalogue of the Exhibition in Palazzo Fortuny, Venice. Venice: Asamer.

Winnicott, D. (1971). *Playing and Reality*. London: Tavistock Publications.

Chapter 7

Climate Change and Adolescence

A Dangerous Collusion of Internal and External Catastrophe

Christine Franckx

Introduction

The end of times has been predicted over and over again, but never in human history have we been so realistically confronted with the possibility of a catastrophic end of the world that is overwhelming all capacity for representation.

We need to facilitate representation of new ideas, new solutions to the old conflicts of struggle between Eros and Thanatos instincts, between sexual and aggressive drives, and assist youngsters taking a place in today's world. The role of psychoanalysis is important in this matter, albeit complex and challenging, to help develop the articulation between the contemporary external and the timeless internal human world.

Adolescence is a phase in human life that brings the process of integrating old and new values to the foreground. It is a time of catastrophic change as a condition for growing up. This chapter will concentrate on the superposition of the adolescent process with climate change and environmental crisis with its catastrophic dimensions (Franckx 2023).

Climate Change, an External Catastrophe in Search for a Story

According to Bertrand Vidal (2018), the announcement and anticipation of a catastrophe can be a way to infuse new meaning to life and prepare for post-apocalyptic times. All sectors of artistic expression, literature, film, pictural art, poetry, etc., provide utopian and dystopian images that meet the human need for representation to prepare for the future. Some thinkers and artists prove in retrospect to be visionary in their prediction of the course of evolution.

Apparently, our deep human desire to give meaning to life and know our place in the universe drives us towards the important questions of science and philosophy. What is time? What is the origin of the universe? What laws do we obey? And what does that say about our existence?

Nevertheless, the anxiety referring to the threat of the ending of the world with the extinction all human and non-human life, has become very real and tangible in our everyday life. Progress is no longer guaranteed and the divide

DOI: 10.4324/9781003498605-7

between the northern and the southern parts of the planet Earth becomes deeper and deeper, with an unequal distribution of riches and of environmental conditions for a safe and prosperous life.

Amitav Ghosh (2021) argues that climate change and the ensuing environmental catastrophe have their origin in the search for nutmeg in the early 16th century with a ruthless exploitation of the natural environment and a full-scale genocide of the population of Banda Islands (Indonesia) as a consequence. For this indigenous people, the beautiful landscape of their islands, with its forests and trees, with its mysterious and feared fire-spitting volcanoes, did not only provide a fertile environment for growing nutmeg, it also was a place full of imaginary and material richness entangled with human life. Western colonialism was based on exploitation of natural goods and the relationship humans could entertain with their natural surroundings was dismissed as superstition, to be eradicated. The centuries-long geopolitical exploitation of the planet that followed, still active today, is based, according to Ghosh (2016), on a mechanistic view of the Earth. The overall exploitation of the nutmeg, which was an extreme luxury at the end of the Middle Ages with a handful of nutmeg the equivalent of the price of a house, can be read as a parable of the current environmental crisis.

Man started to see himself as the ruler of Earth and Nature with all its creatures, that had no other reason of existence then to be available as a resource for humans to use for their own needs. The connection between life giving force to the planet and the history of mankind became weaker as the earth became to viewed as "an enormous machine made of inert particles, constantly in ceaseless motion."

The world changed dramatically when the Western, civilized world set out to conquer other civilizations and globally imposed Galileo's image of the world as composed of "objects that were placed next to each other, without mutual influence."

The Enlightenment of the 18th-century movement applied optimism, empiricism, and rationalism of the scientific revolution to society and placed man at the top of the natural pyramid and as the only living creature capable of communicating, thinking and therefore responsibly acting. The taxonomic category of *Homo sapiens* dates from 1758, which is coincidental to the development of sciences against a cartesian background of *Cogito, ergo sum*.

Today, according to Bruno Latour (2015), we are living more in a world conceived of as Gaia, a world of actors who are constantly interacting with each other in an unpredictable way. The experience of life on earth is the secondary consequence of the interaction of all living organisms.

A new climate regime, some called it the Anthropocene, invites to reconsider the place of mankind on a planetary scale. Economists, historians, legal experts, technologists, sociologists, and all branches of science are impacted by the now broadly accepted evidence of a threat to the existence of our species, and in fact all life on earth.

Global warming is setting in motion a highly complex process of physical, biological, social, and psychological interconnections.

The Gaia-hypothesis, named after the ancient Greek Goddess, is a model of the earth as a complex interacting system of all living and inanimate organisms, that sustains life through homeostatic regulation. It was developed in 1974 by James Lovelock and Lynn Margulis and remains an interesting candidate to understand the fragile living conditions on the planet.

For a long time, attentive observers have noticed the importance of an intimate relationship between human and non-human life and have tried to understand when and how this connection loosened and started disappearing.

The French author Elisée Reclus, who lived from 1830 to 1905, clearly understood the intimate link between the environment and the human psyche when he wrote in *Du Sentiment de la Nature dans les Sociétés Modernes* (1866): "that a secret harmony is established between the earth and the peoples it nourishes, and when imprudent societies allow themselves to lay hands on what makes up the beauty of their domain, they always end up repenting. Where the soil has become ugly, where poetry has disappeared from the language, imaginations are extinguished, minds are impoverished, routine and servility take hold of souls and dispose them to languish and death."

Historians describe new ways to think more globally about historical regimes at the crossroads of ecology and the world. The Bengali historian Dipesh Chakrabarty (2021) was the first to apply history-writing to the planetary age, essentially taking the temporalities of the world together with those of the Earth system with a change of scale. The age of the cosmos is approximately 13.5 billion years, of the earth roughly 3.2 billion years and of *homo sapiens*, 300,000 years ago, which is yesterday on a planetary scale. He affirms that humanity has become a geophysical force that has the power to impact the history of the earth, which implies that we can no longer separate natural from human history. Climate warming is transforming history of life and is in itself provoking a sixth massive extinction. Nature is no longer a background object without internal intentions that changes the course of human history through natural catastrophes (such as earthquakes, epidemics, or volcanic eruptions) but is itself being influenced by human activity on a large scale. This implies a radically different historiography that situates man and nature in a reciprocally interactive relationship on a much larger time-scale (Chakrabarty, Haber, and Guillibert 2017).

From a legal perspective, Christopher Stone argued already in 1972 that natural objects should have legal rights. In 2017 the New Zealand government granted the river Whanganui personhood status and declared it a living whole, from the mountains to the sea, incorporating all its physical and metaphysical elements. This legal act was a precedent that has been followed by some other countries, including Bangladesh, which in 2019 granted all its rivers the same rights as people. The European Parliament adopted in 2019 a resolution declaring a climate emergency for Europe and the world[1].

These are just a few examples of the way climate change and its dangerous catastrophic potential are writing the story of humans.

Adolescence, an Internal Catastrophic Change

At the time of the conquest of the world, which ended the Middle Ages and opened the world on many levels, the concept of "adolescence" did not exist, nor was there any notion of "childhood." Children were mini-adults and youngsters were valued for their strength, impetuosity and fearlessness, in short, in times of war, as "cannon fodder."

It wasn't until the end of the 18th century that this period was considered to be specific to the psychological development of children into adults. In his *Emile* (1762), Jean-Jacques Rousseau (1712–78) advocated an education in line with the child's natural development, whereas until then training, or pedagogy as we would say today, had been the main means of guiding the child towards maturity, of "making" the child an adult in a way.

The discovery of the psychic nature of adolescence as a particular pivotal phase between childhood and adulthood, which defines our vision of the human being and differentiates us from other mammals, cannot be overstated.

Two major changes begin to take place: the body matures sexually and childhood objects are released. A complex process of identification and dis-identification then begins, with a recapitulation of the first stages of development, but now "big," and "for real." This is a period during which the ability to mourn and to allow for feelings of ambivalence and dependence are launched with momentum.

The Drives: A Story of Fire and Flooding

The drive is defined as a boundary term between the psychic and somatic and is a measure of the work the psyche has to do to bind the impulses coming from the body.

We distinguish puberty, which is the period that starts with hormonal maturation and thus initiates a chain of biological changes and a growth spurt of all somatic systems, also affecting the development of the brain: a cognitive leap will bring concrete to formal logical thinking, and relationships within a complex problem can now be contemplated in a systematic way. This is a not insignificant factor in understanding this stormy evolution (Gutton 2013).

The term "adolescence" refers to the period of psychic adaptation to this new reality and this continues into young adulthood, nowadays referred to as the "transition age," with taking action and realizing personal projects.

In the *Three essays on the Theory of Sexuality* (1905), Freud needed 15 years and several editions to integrate the crucial role of adolescence into his views on the role of infantile sexuality in adult object choice. In the evolution from infant to adolescent, psychic reality takes shape in the contact with the drives in two times, interrupted by a latency phase of waiting for maturation and for the release of the first love objects. Freud thought of latency as a typically human fact, and one of the causes by which man can develop a higher culture, but also runs the risk of becoming neurotic. It is a fallacy to think that Freud saw the child as the object of psychoanalysis, he was concerned with the infantile as a kind of myth that lives in everyone, regardless of age. But there is, of course, always the factor of development, which is age-related, and in addition, the external reality that defines us all. Identification brings drives and objects together: after all, the world was there before the infant was born and he is helped by his environment to take it in progressively so that the inner world acquires a geometry that refers to the external.

It is the puzzled question of someone looking at a work of art: do I like this because I already know it well and recognize it, or am I surprised by something beautiful that now enriches my internal world? Or, it is also heard to say: in nature, "ugly" does not exist! Is this because nature belongs to us and it is intuitively recognized as such? There is always a connection between inside and outside, projective and introjective identificatory processes are constantly at work.

Among mammals, humans are an exception: babies are born prematurely and are unable to go in search of food; they need others for their survival. But the main function of these "others," the first objects, is to maintain the illusion in the baby that instinctive needs can be satisfied by him alone. If all goes well, the baby is reinforced in this delusional perception of being fully in control of the world: he hallucinates the breast when it is not there, he shuts himself off to too much excitement in a negative hallucination movement by withdrawing into himself. In short, babies use positive and negative hallucination as a way of imagining themselves in complete control of events. Everything is dominated by the pleasure principle and the infant approaches the world as best he can, in other words, in small steps and only in small fragments of the whole. In fact, the baby's ego is immature and will only develop gradually and piecemeal. Biology determines the rate of development at which external reality and the internal world can take shape simultaneously. A first *good enough* object will make the delicate work of disillusionment begin at the right moment - the baby must be helped to assimilate the impact of the reality principle little by little. You could say that the drives are a way of giving instincts room to breathe and providing the psyche with something to "think," to imagine, to fantasize about.

Why talk about babies now? Because the same psychological mechanisms will reappear in adolescence. The reasonableness of latency is lost, and those who have themselves helped children through this period will surely recognize how suddenly a greater attention and presence were over again required to keep up with the new unknown and disruptive.

The proximity of the psychic world of infants and of adolescents, albeit in a different developmental context, demonstrate the challenge of the intensive psychic work adolescents have to go through. Would we consider our planet to be one big dumping ground, in a primitive reflex of the abused infant who, from a position of dependence, pours out his rage, fury and hatred in his mother's imaginary womb, as if it were a "toilet womb" on a planetary scale where he could dispose of all his emotions and waste? After all, the infantile feelings of dependence on his mother are painful and he hates her for it, he would like to punish her, control her and destroy her completely inside. It's a pleasurable destruction that may give short-term satisfaction because of the relief, but it is followed by guilt and a reparation movement with the same motive: the baby needs its mother.

In adolescence, the theories the young child developed in his questioning of the mysteries of life, "where did I come from," and "where is life taking me," "who are my parents," and 'who am I," […] are put to a new scrutiny.

Mourning Safe Beacons

The early psychological processes that balanced growth towards a degree of independence and autonomy will now be the subject of a new edition.

In this way, the relationship with the external and internal objects of childhood is fundamentally altered: the adolescent forms his or her own personality and prepares to take his or her place in society.

We read in Freud (1905, 227):

> "At the same time as these plainly incestuous phantasies are overcome and repudiated, one of the most significant, but also one of the most painful, psychical achievements of the pubertal period is completed: detachment from parental authority, a process that alone makes possible the opposition, which is so important for the progress of civilization, between the new generation and the old."

The relationship between parents and children suddenly changes and it's as if, for both parties, all the resources they had previously trusted were taken away from them. The Italian writer Erri De Luca puts it this way in his book *A Grandezza Naturale* (2021), which contains stories about the relationship between children and parents and says that the no-man's-land of adolescence suddenly emerges between parents and children. Adults are the past. Their voices of reproach, even when shouted, are muffled. This desert must disappear on its own, no direction can shorten it. Bereavements reinforce it, first loves prolong it.

An essential loneliness speaks from this description, being without any direction and broken connections without the hope of finding the way out of this desert. It is a time of mourning and breaking free from initial object relations, the road must be travelled alone, there are no signposts.

It is also a time when the Eros of connection and the Thanatos of dis-connection and dissolution, must find new cooperation. Old patterns have to be let go, and new elements come together in a different arrangement. The death drive comes to the fore in different ways: as a desire to stop time and dwell in a garden of Eden of paradisiacal happiness for eternity; we think, for example, of the anxiety that sometimes becomes more prevalent as the end of studies approaches, but also the sense of a rushing forward of time that the young try to keep up with by taking action and smashing what provided peace and stability. Fortunately, Eros is also more strongly present as adult life beckons, the desire to participate is intense and there is now also the prospect of committing to a sexual partner and putting the terror of finality of life at bay by caring for the next generation.

In order to grasp the inner turmoil and reconcile it with the demands of external reality, teenagers also need sufficient freedom to explore and oppose authority in powerful political and social ideals. More specifically regarding the ecological crisis, we see how this mobilizes young people worldwide, because this concerns them, but also how this engagement simultaneously risks not finding the right parental and social framing needed to outgrow the illusion of infantile omnipotence. After all, there is a real threat to humanity. "Saving the Planet" may keep them stuck in unresolved inner identity con-flicts because the external objects, so much needed as solid and reliable models of identification in the background, are themselves in crisis.

Not only are adolescents anxious about the uncertain future of the envir-onment, but they may also feel anxiously haunted by their primary objects that have failed to adequately protect them. This new global geo-socio-poli-tical emergency affects the development of their ego functions with a risk to a fragile constitution of their adult selves.

Today, society as a whole is going through a phase of transition, with the loss of familiar values and a search for new references that are not yet stable or reliable. There is a general climate of profound uncertainty and unclear expectations, accentuated by rapid technological change. The more fragile egos of teenagers are vulnerable without the presence of support in the background to help manage the excitement caused by new contact with events in society. The problem is that the internal emotional turbulence and the search for safe and reliable reference points in adolescence are reproduced exactly in today's society.

German sociologist Hartmut Rosa (2016) explains in his work how the image of the adult as an authoritarian educator has been transformed into that of a partner willing to cooperate, and how this also has a dark side: adults too are caught up in an increasingly intense acceleration aimed at undoing the finitude of existence. They give way more harshly to youth, for example mothers seeking the appearance of their daughters, fathers wearing the same clothes as their sons. In itself, this is nice and playful, but it can make it harder for teenagers to move forward, to commit the symbolic father murder. We have long observed that the vertical and hierarchical dimension of social structures has weakened or even disappeared in favor of a more horizontal and fraternal dimension. The period of adolescence continues into young adulthood or the age of transition, and the change of generation is more complicated. We can see the effects more clearly in the following two areas: global warming and the growing influence of social media. In the first case, teenagers are reminding adults of their responsibilities and are angry at the neglect and missed opportunities to save the planet by previous generations. In the second case, teenagers are virtually left to their own devices when they are "connected" via social media without adequate and expert guidance from their parents. One of the signs, some will call it symptoms, of the changes in the social contract between generations of our changing time is the effect on language structure and uniformization of vocabulary with a loss of articulation and differentiation (Lebon 2022).

Development or Catastrophe

The experience of time and the subjective feeling of existing in time are related to the working through of mourning, an important developmental aspect of adolescence.

Freud wrote the essay "On Transience" in the same year as his metapsychological paper "Mourning and Melancholia," showing the psychological importance of being able to integrate the pleasure and reality principles:

> "No! it is impossible that all this loveliness of Nature and Art, of the world of our sensations and of the world outside, will really fade away into nothing. It would be too senseless and too presumptuous to believe it. Somehow or other this loveliness must be able to persist and to escape all the powers of destruction. But this demand for immortality is a product of our wishes too unmistakable to lay claim to reality: what is painful may none the less be true."
>
> (Freud 1916, 305)

David Bell (2006, 783) argues that "feeling oneself as existing in time is an important developmental achievement. For some, however, it is felt as a fixed, imminent catastrophe to be evaded by the creation of a timeless world in which, apparently, nothing ever changes, where there is an illusion of time standing still."

The novel by Jerome David Salinger of 1951, *The Catcher in the Rye* is a very good illustration of the instability and turbulent emotions involved in looking for a good balance between the blessings of a protected childhood and the demands of adulthood. It tells the story of two days in the life of a 16-year-old boy, Holden Caulfield, who is expelled from school and wanders alone in New York City. He feels trapped in "the other" side of life, childhood, and he cannot find his way in a world that feels phony and alien to him.

"Certain things, they should stay the way they are. You ought to be able to stick them in one of those big glass cases and just leave them alone" (p. 120).

The boy eventually ends up in a mental hospital because of a psychic breakdown and this quote refers to his troubled state of mind. Nevertheless, the story movingly describes his passionate search for belonging to a world with adult people he can try to identify with, while at the same time he is being pushed forwards and away from his childhood identificatory figures, who are no longer of use to him as his parents don't even know he has been expelled from school. The book has become an icon of teenage rebellion, but also feared by educators for the sexual and societal liberal themes that could spoil young people.

Global Warming and Saving the Planet!

Global warming is undoubtedly affecting this development process. Will internal objects survive the adolescent's attacks, and can the previous generation confidently represent a process of global transformation of the outside world? The anger of adolescents reveals the violence to which they feel exposed as they discover not only the scientific facts of global warming and the disappearance of many species, but even more so as they realize the inability of their parents' and grandparents' generations to respond appropriately. The childhood fear of feeling totally dependent on the primary/maternal object for survival is reactivated in the normal individuation process of adolescence but is now further intensified and compromised by the realization that Mother Earth has become vulnerable and that the climate may no longer provide a safe home in the future.

In human relationships we can expect a certain degree of interdependence and agency, but Nature is utterly ruthless and goes its own way without any intention. She doesn't need us to survive. For the first time in human history, the fear of collapse is projected into the future and is not something that has already happened but has not yet been internalized, according to Winnicott (1974). This affects the adolescent's existential questions and his or her position in the chain of generations.

We are living today in a very complex situation that raises questions about relations between the generations, the role of the individual in the group and the social contract between people. Enormous ethical demands are being placed on the next generation, with no adult authority to point the way from their own experience. As one intelligent 14-year-old girl said to her mother: "We are the first generation to be really confronted with the climate crisis and we are also the last to be able to do anything about it."

"Saving the planet!" Indeed, faced with this global catastrophe, there seem to be three possible ways to react: denying reality and carry on as usual; protesting out of a sense of injustice and anger or accepting the real threat that inevitably brings fear and psychological suffering. These reactions are universal and occur in everyone at different times: usually we need some denial and disavowal as a defense in a functional splitting to protect a vital sense of *going-on-being*; at other times we may be very angry and defiant, but we all experience moments of anxiety.

The climate crisis is infiltrating the development process of adolescents. It invades their existential questioning and impacts on their transgenerational relationships. In recent years, we have witnessed many forms of ecological activism on the part of teenagers committed to Saving the Planet, who have no choice but to do otherwise. They passionately aspire to a reliable and promising future to fuel their ideals and projects, while at the same time they are confronted with the urgency of their individual pubertal crisis in their private natural environment (Bernateau 2021). This concomitance puts the adolescent mind in the firing line!

Adolescents show impressive lucidity on this issue and paradoxically the reality principle seems to be more accessible to the younger generation. They are better informed and ready to take action because of their sensitivity to what is going on in tomorrow's world, their world, and their eagerness to get involved. So, it is no surprise that the ecological crisis is mobilizing adolescents worldwide in "Saving the Planet," but this, in my opinion, brings two additional risks to the adolescent process.

First, there is the risk of not finding the right parental and social embeddedness needed to transform their illusion of infantile omnipotence because of the real threat to survival. This can keep adolescents trapped in a state of constantly repairing the damaged parental objects rather than adopting a depressive mourning position.

The fundamental role of parents (and their substitutes) is to provide to the next generation a containing of protection and sense-making. Climate warming and the global crisis have weakened this function. How can parents of today can help the younger generation to imagine and mentally construct a solid project for the future in a world that has become ephemeral? (Robin 2023).

Second, the internal and libidinal crisis of catastrophic change in adolescence can be overtaken by the truly catastrophic external environmental threat, and take over the whole world for adolescents, who then close this necessary and vital intrapsychic stage in their development. Any internal movement and drive can be potentially traumatic for the adolescent if the psyche is unable to sufficiently integrate the life and death drives.

The condensation of the two times in the libidinal elaboration of the infantile sexual theories, not allowing a "latency" time, thrusts the child into adulthood and prevents that the climate threat acts on the adolescent according to the catastrophic mode of a traumatic neurosis (Glas 2022).

This author underlines the danger for adolescents, focusing on saving the planet, of not being able to construct a psychic organization under the reign of a super -ego organized by oedipal conflict (ibid.).

Clinical Example

Jennifer, aged 15, was referred because of a serious suicide attempt that her parents and friends did not see coming. They were shocked to realize that this perfectly happy girl, always a good student and her mother's confidante, had suddenly collapsed for seemingly no "reason." In an initial conversation, Jennifer reveals how exhausted she has become by her efforts to meet all the demands of her parents and teachers, to project a glowing image to the outside world, while internally she was extremely gloomy and preoccupied. Feeling that there was no future for her in this world, she did not find a way to be happy. Lonely and isolated from her peers, she compulsively seeks out fulfilling contacts through social media. However, this only adds to her disappointment. In one of the psychoanalytic sessions that follow, Jennifer reveals how deeply depressed and hopeless she feels. The psychoanalytical work is laborious and the internal conflicts are slow to emerge. At one point, she tells me how painfully disappointed she is in her mother, with whom she had felt very close as a child, now that she has discovered her mother's extramarital affair. It then became apparent that Jennifer had always felt more like her mother's partner, to compensate for the emotional coldness between her parents, and that she has now been replaced by a real lover. What is most painful for Jennifer is that she suddenly woke up from a state of "illusion" and reality hit her in the face. She feels cheated on and betrayed. At last, she is able to express her anger and sense of injustice. She then has a dream in which she is in a shopping center with a young woman she doesn't know, when they suddenly find themselves trapped inside the building, which catches fire. She can't get out and feels an unbearable suffocating heat. What's more, when the fire brigade finally arrives, they flood the building without noticing the people inside. She loses the other woman's hand, almost drowns and wakes up in a panic. Although this dream seems to indicate repressed homosexual feelings, awakened in transference, and her fear of being literally submerged and almost drowned in the intense libidinal experience, her immediate associations refer to the effects of global warming. I interpreted this as a resistance to engaging with her internal arousal and formulated something about the fear of being overwhelmed in close relationships, but she didn't accept this and got angry with me for not noticing the reference to the environment. It turns out that she is heavily involved in a social media group that focuses on climate activism. There is no choice at this stage but to take her seriously on the issue of the environment and accept her understanding of her dream as a constructive and sincere attempt to participate in adult life. In the counter-transference, I realize that the images in her dream had made me very anxious and that my interpretation at the oedipal libidinal level has served a welcome escape for me, relying on my psychoanalytic profession.

This example made me think how important it is today, in analytic work with adolescents, to properly understand the superposition of eco-anxiety and the anxiety caused by neurotic conflict, as a symptom of our contemporary collective suffering. First interpreting our shared anxiety and helplessness about the bleak aspects of the future before incorporating the dissymmetrical transferential meaning of the analytic material, as in Jennifer's dream, is often necessary and can further pave the way for more authentic and necessary intrapsychic work.

We hear young people doubting whether they want children, whether they will eventually die of old age, and so on, and this may obscure some internal hesitation about assuming an adult position in society. Models of identification have become fragile. Yet it is only by mobilizing internal fantasy life and symbolic capacity that a real life becomes possible. Though more complicated in our time, helping adolescents achieve this mature psychic state is a necessity for their future.

The Dangerous Collusion of Internal and External Catastrophe

The difficulty in our contemporary society is that internal representations of life and death, of reproduction and attachment, of all attacks on linking, are being reproduced in the external world.

"The ice cracks violently, water gushes between the colliding debris, but yet time passes in slow motion. An elk appears in our field of vision, perched on a drifting iceberg. He must have been in the middle of the river when it happened, when the frozen river suddenly broke up beneath his feet, everything went very fast then, everything started to move, he was dragged along. He passes about twenty meters from us, stuck on his piece of ice as it sails downstream, his eyes haggard, he must be scared. The scene lasts a few seconds, the moose and its iceberg disappear as they appeared behind an arm of the river. We look at each other. "That's it," says one of the hunters. That's a bit like us. We're being swept along by a current we can't control, the ground has opened up beneath our feet, there's nothing solid left to hold us. We don't know where we're going anymore" (Martin 2016, 31).

Anthropologist Nastassja Martin has given a moving account of her experiences in the Arctic and sub-Arctic regions, including Kamchatka, a peninsula in Russia's far east bordering the Bering Sea, and Alaska.

For the inhabitants of the Far North, dreams are of the utmost importance and are a privileged means of maintaining vital contact with nature and 'other' animals in these arid and isolated regions. It is a way of receiving an infusion, as it were, of the environment and of learning, through dream life, about migrating herds of animals and approaching predators. The images that emerge from nocturnal dream life bear the traces of daytime impressions and are considered to be precise clues, guiding decisions as a matter of survival.

The fact that silence and apathy have prevailed around this enormous problem, even in psychoanalytic theoretical and clinical work, shows the power of the unconscious destructive forces at work here, which have allowed a signal-anxiety to escape the radar, to remain out of consciousness.

The notion of "Catastrophic change" was introduced by Bion in 1966 in a lecture given to the British Society of Psychoanalysis. The text of this lecture is reproduced in *Attention and Interpretation* (1970) and is relating to a psychotic patient. Bion wrote that mental evolution or growth is catastrophic and timeless and that catastrophic change may occur each time a new thought enters, but he did not expand his assertion to imply that all emotional growth and change is at some level experienced as catastrophic and threatening.

Other authors have elaborated on the meaning in a developmental context in an attempt to continue in Bion's line of a theory of thinking. Howard Levine (2009, 77) proposes that Bion may have referred to Freud's writings (1920, Beyond the Pleasure Principle; 1923, The Ego and the Id), and that "catastrophic change" may "recognize that reality contains far more than ever can be represented [...] and that psychic change and emotional growth require an encounter with that which is truly new and unknown, the latter with a status of inchoate, chaotic, disorganized and potentially disorganizing beta-elements, to be contained and transformed before new representations can be created. In such circumstances, change and growth may be seen as requiring an encounter with the unknown that may well feel or even loom as catastrophic and destabilizing."

Further development of a concept that has a potential to contribute to the evolution of psychoanalytic thinking and practice can amongst other be found in an article of Renata Gaddini (1981), according to whom both Bion and Winnicott thought that Freud have directed his psychoanalytic investigations too exclusively to the repressed. Winnicott writes of the fear of breakdown, and Bion writes of catastrophic change. Both focus on the dread of encountering emotional truth encapsulated in the unrepressed unconscious, threatening the mind with a psychotic state. Yet both contend that this encounter can save the personality from psychic death as a defense against it. Catastrophic change is made possible through an encounter with the unrepressed unconscious in the present, in the transference, primarily through the analyst's capacity to experience the agonies of breakdown in his own flesh. It is the analyst who must "agree" to experience a catastrophic change, to lose his identity, hence enabling the patient to dare to approach emotional truth. Both the "fear of breakdown" as well as the "fear of catastrophic change" are the equivalents of the fear with which one meets, in the course of the analysis, and these hold back the patient, desperately cohesive fearful of changing, as if he were on the very edge of a catastrophe or a precipice (Gaddini, 1981).

Not only are adolescents pessimistic about the uncertain future of the environment, they may also approach the emotional truth of their primary objects failing to protect them sufficiently and feel anxiously persecuted by them. This new global geo-socio-political emergency situation therefore impacts the development of their ego-functions with a risk for a fragile constitution of their adult self.

The world that previous generations created and lived in has dramatically changed and is no longer solid enough to be challenged, while the future is equally uncertain. For the first time in human history, the fear of breakdown (Winnicott 1974) is projected in the future and is not something that has already taken place but was not internalized. In the current situation of climate change, there simply is no representation at all.

Because there is no object involved to possibly identify with, we are in a very different asymmetric relationship from the human dependency on another person for survival and growth. This complicates the adolescent process as the real adult world with its many challenges and opportunities that confront young people now includes for the first time in history a reality without an object to challenge.

The catastrophe of mental growth with its characteristics of deconstruction of an earlier state and the encounter with a new reality needs the other to overcome, whereas the external catastrophe is attacking the function of the other in a very profound way that has not been seen before.

The Greta Thunberg Generation

When this teenager addressed the United Nations with the accusatory words "How dare you!" or "blah, blah, blah," putting her finger on the sore spot of the inactivity of the men in power, we hear an Oedipal tragedy: is the child sacrificed by the parents so that they can continue to live in peace and tranquility, just as Laius and Jocasta left their son Oedipus for dead in order to escape the oracle's prediction that their child would kill them? Luc Magnenat (2019) develops the hypothesis of biocide as infanticide in his edited book *La Crise Environnementale sur le Divan*.

Magnenat suggests that Greta Thunberg can be understood as Antigone who wants to restore the order violated by the previous generation. Indeed, the climate crisis is reminiscent of the Oedipal myth, according to which the biological parents of Oedipus tried to kill their son in the vain hope of being delivered from the impending prophecy that he would one day be murdered by him. Did the previous generation sacrifice its future offspring to remain, as Sally Weintrobe (2021) put it, in a bubble of denial and magical omnipotent thinking to rearrange reality with fraudulent arguments.

The generation of Thunberg underlines in their different actions the unreliability of previous generations who not only created the catastrophe but were incapable and unwilling of taking appropriate measures to limit the damage. It's as if the world is working in reverse: rather than teenagers being reprimanded by their parents for their irresponsible behavior, young people are now reminding adults that they must rein in their insatiable infantile impulses of greedy orality and face the facts.

The anger of the teenagers reveals the violence they are suffering as they discover not only the scientific facts of global warming and the disappearance of a large number of animal species, but even more so as they realize the inability of their parents and grandparents to react appropriately.

Conclusion

The climate crisis is calling into question intergenerational relations, the place of the individual and the group, the social contract between people, the value of exploitation and sacrifice, and the importance of collaboration and solidarity (Desveaux 2020).

As well as being trapped in unresolved identity conflicts, some teenagers also run the risk of "Saving the Planet" becoming their main libidinal goal during the narrow window of adolescence. The intimate evolution of puberty could then be displaced towards involvement in an undifferentiated group of adolescents around the world supported by a more noble and idealistic goal rather than struggling with their own sexuality. Fragile adolescents may run the risk of finding an easy way out of internal transformations that seem unmanageable. Their difficulty in preparing for libidinal satisfaction with an appropriate sexual object may be overlooked as they become enamored with eco-activism (cool kids saving a hot planet), rather than engaging with their own private urges.

As a result, the struggle against the destructive forces of the external environment can overlap and mask the intrapsychic and fantastical struggle with the first internal objects, the primary goal of adolescence. The scientific reality of global warming and its potential consequences for the survival of the planet can bring adolescents into contact with the fear of collapse which, according to Winnicott (1974), refers to a catastrophe that has already occurred but remains unresolved intrapsychically. It can cast a shadow over intimate reality and, in a way, steal the internal conflictuality of the adolescent's ambivalence, so necessary for growing up and/or recovering from archaic anxieties.

The "How dare you?" addressed to adults takes on its full meaning if we consider the psychic damage suffered by adolescents who are not allowed to experience a creative process specific to adolescence!

In taking the anxiety about climate change seriously we can find a way to connect with the adolescent that conveys a shared collective feeling about the uncertainty of the future, while at the same time maintaining the intrapsychic conflict and transference dimension in tension. This way we offer a reliable identification figure, strengthen ego-functions in the libidinal chaos of adolescence and help the youngster in reconnecting with good internal objects. After all, they are the future and will need a healthy psychic structure to deal with the new problems of humankind.

Note

1 www.europarl.europa.eu/news/en/pressroom/20191121IPR67110/the-european-parl
iament-declares-climate-emergency.

References

Bell, D. (2006). "Existence in time: Development and catastrophe". *The Psychoanalytic Quarterly*, 75: 783–805.

Bernateau, I. (2021). "Menace sur la Terre et vulnérabilité adolescente". *Adolescence*, 39(1): 31–42.

Bion, W.R. (1970). *Attention and Interpretation*. London & New York: Routledge, 1984.

Chakrabarty, D. (2021). *The Climate of History in a Planetary Age*. Chicago, IL: University of Chicago Press.

Chakrabarty, D., Haber, S., & Guillibert, P. (2017). "Réécrire l'histoire depuis l'Anthropocène". *Actuel Marx*, 61(1): 95–105.

De Luca, E. (2021). *A Grandezza Naturale*. Milan: Feltrinelli.

Desveaux, J.-B. (2020). "La crainte de l'effondrement climatique. Angoisses écologiques et incidences sur la psyché individuelle". *Le Coq-Héron*, 242(3): 108–115.

Franckx, C. (2023). "Malaise dans la Nature. La psychanalyse face à la crise climatique". *Revue Belge de Psychanalyse*, 82(1): 111–136.

Freud, S. (1905). Three Essays on the Theory of Sexuality. *SE* 7.

Freud, S. (1916). On Transience. *SE* 14.

Gaddini, R. (1981). "Bion's 'catastrophic change' and Winnicott's 'breakdown'". *Rivista di Psicoanalisi*, 27(3–4):610–621.

Ghosh, A. (2016). *The Great Derangement: Climate Change and the Unthinkable*. London: Penguin Books.

Ghosh, A. (2021). *The Nutmeg Curse: Parables for a Planet in Crisis*. Chicago, IL: University of Chicago Press.

Glas, J. (2022). "L'investissement massif de la décroissance écologique, un avatar inconscient de la jouissance dans le dépouillement de la castration". In D. Bourdin & D. Tabone-Weil (Eds), *Planète en Détresse. Fantasmes et Réalités* (pp. 23–36). Paris: PUF.

Gutton, Ph. (2013). *Le Pubertaire*. Paris: PUF.

Latour, B. (2015). *Facing Gaia: Eight Lectures on the New Climate Regime*. C. Porter (Trans). Cambridge, MA: Polity Press, 2017.

Lebon, C. (2022). "L'écologie familiale ou la famille à l'épreuve de la crise écologique mondial". *Divan Familial*, 49(2): 69–83.

Levine, H.B. (2009). "Reflections on catastrophic change". *International Forum of Psychoanalysis*, 18: 77–81.

Lovelock, J.E. & Margulis, L. (1974) "Atmospheric homeostasis by and for the biosphere: The Gaia-hypothesis". *Tellus*, 26: 1–10.

Magnenat, L. (Ed). (2019). *La Crise Environnementale sur le Divan*. Paris: In press.

Martin, N. (2016). *Les Âmes Sauvages. Face à l'Occident, la Résistance d'un Peuple d'Alaska*. Paris: La Découverte.

Reclus, E. (1866). *Du Sentiment de la Nature dans les Sociétés Modernes*. Paris: Bartillat, 2019.

Robin, M. (2021). "How dare you? La jeunesse en mode survie". *Adolescence*, 39(1): 15–30.

Robin, M. (2023). "Sacrifier l'enfant pour sauver le contenant". *Carnet Psy*, 264: 30–32.

Rosa, H. (2016). *Resonance: A Sociology of Our Relationship to the World*. J.C. Wagner (Trans). Cambridge, MA: Polity Press, 2021.

Salinger, J.D. (1951). *The Catcher in the Rye*. New York: Little, Brown and Company.

Stone, Ch. (1972). "Should trees have standing? Towards legal rights for natural objects". *Southern California Law Review*, 45: 450–501.

Vidal, B. (2018). *Survivalisme. Êtes-Vous Prêts pour la Fin du Monde?*Paris: Arkhê.

Weintrobe, S. (2021). *Psychological Roots of the Climate Crisis*. London: Bloomsbury.

Winnicott, D.W. (1974). "Fear of breakdown". *International Review of Psychoanalysis*, 1: 103–107.

Chapter 8

Birth is not Destiny

Orazio Pietro Attanasio

Introduction

A possible interpretation of catastrophism is that humans are not in full control of their own destiny, and that, at least after some point in life, subsequent developments cannot be altered in a substantial fashion. Of course, that might be particularly despairing for those who have drawn a short straw in the lottery of life and elating for the lucky ones. However, from a societal point of view, or behind the veil of ignorance, before one knows its own destiny, such a society is a dystopic one.

We know that the features that characterize human abilities in various dimensions have much persistence and what happens in the early years of life, the first thousand days since conception, has long-term consequences that are then visible in a variety of outcomes. Many of the events that might imply good or bad life-cycle luck cannot be controlled. At conception, different individuals are endowed with different combinations of genetic material. No one can choose their own parents or even what genetic features they inherit from their mother or their father. However, immediately after birth, the phenotypes generated by these combinations are exposed to a variety of environmental factors that shape their expression and development. Parenting practices, the availability of social support, the quality of nurseries, schools and peers, the influence of neighborhoods all contribute to determine in an important fashion, as time goes by, life outcomes, well-being, happiness.

It is therefore clear that human development is not completely fixed and determined at birth. Indeed, early and subsequent developments are very malleable. In this sense, the old debate between the influences of nature and nurture in human development is exceedingly restrictive. And this is important: recognizing that early childhood development can be affected by the environment where children grow up implicitly recognizes that initial disadvantaged can be compensated. More importantly, the fact that early disadvantage, while not being controlled by the individuals that experience it (the children), can be changed, draws attention to the moral necessity of designing policies that can allow the full development of every individual.

DOI: 10.4324/9781003498605-8

From Birth to Adulthood

Consider two children born in the same year, in the same city, but a few miles apart. The first is born into a poor neighborhood, to parents facing many financial problems, with relatively low education attainment. The second is born in a much more affluent neighborhood, to parents who are much better off economically, with well-paid and stable jobs, which they could obtain in part because of their high level of education. If these two children and their development are observed at age three, five, or later, such as at ten or 16, it is easy to predict that the first child will lag behind the second, in terms of cognitive and socioemotional skills. These differences emerge early in life, and they tend to become larger. They are then reflected into different academic attainments, different propensity to engage in criminal activities, in the type of jobs that they can obtain in the labor market and the type of family they will build (or not).

Obviously, the degree to which certain types of skills and activities are remunerated during adulthoods will depend on the type of technology that is prevalent in the production of goods and services and on the availability of different skills in society. However, it is clear that a substantial part of the differences in outcomes in adulthood relevant for individual well-being are determined by the level of skills with which individuals are endowed as an adult. And these, in turn, are well-predicted by development in the early years. Indeed, even characteristics of individuals at birth (like their weight) seem to be predictive of adulthood outcomes (Almond and Mazumder 2013).

Can we conclude that the *destiny* the two children mentioned above face at birth is, to a large extent, already determined? Do the initial conditions individuals are born with determine the rest of their lives? Is birth destiny?

An answer to these questions is difficult because it is hard to establish strong causal links between the various factors that influence human development. For instance, the environment, in turn, modifies the genotype in a mutual and complex two-way relationship. Even if one does not consider epigenetic effects, which are not uncontroversial, economic factors are likely to affect the way couples are formed (what economists call the marriage market), changing the degree of assortative mating and therefore the genetic composition of the next generation.

Over the years many wrong answers have been provided to the questions above. In the early 20th century, a substantial number of scientists, influenced by research in genetics, seemed to believe that intelligence, as measured by certain IQ tests, is completely determined by the combination of genes a person is endowed with at birth. And for many years, it was hard to remove this belief, as documented in the remarkable story told in the remarkable book the *The Orphans of Davenport*, by Marylin Brookwood (2021).

For anybody holding such a view, the destiny of the two children mentioned above is set at birth. Their respective parents, with whom they share a genetic background, might be poor or well off because of their innate abilities and characteristics, a condition they transmit to their offspring. And the inevitable destiny of society is one characterized by ingrained inequality, which can become even more visible with economic growth if certain skills acquire more importance in the production process and where marriages are increasingly driven by assortative mating. A dystopian and catastrophic view indeed.

The Future Does not Need to be Bleak

Fortunately, such a deterministic view of what determines individual abilities has been repeatedly proven to be unfunded. Genes and genetic background are certainly important. But these factors interact with other many environmental influences that ultimately determine individual outcomes and lives. What is also quite clear now, is that what happens during the early years, has long-term consequences. At the same time, many studies have shown that the early years are malleable.

The history of these studies is an interesting one. One of its first pieces is the remarkable evidence assembled over many years by Harold Skeels, Marie Skodak, and their collaborators who, as documented and beautifully narrated in Brookwood's *The Orphans of Davenport,* noticed improvements in the IQ of "retarded orphans" who were removed from an inhumane environment and sent to an institution for the "feebleminded" to be cared for by "moron" women. The establishment of direct relationships and stimulation seemed to have remarkable impacts in the short and long term. That evidence and subsequent studies effectively debunked the previously held views on intelligence and its formation.

From Evidence to Policy Interventions

While these studies were conducted in the 1930s and encountered remarkable resistance from the scientific establishment, in subsequent decades the tide turned. In the1960s, in Ypsilanti, Michigan, a famous intervention, known as the Perry Pre-School Program was started with the intention to help, very early on, children growing in disadvantaged environments. The Perry Pre-School Program was subsequently evaluated with a rigorous evaluation of their impact in the short and long term (see, for instance, Heckman, Moon, et al. [2010] and Heckman, Pinto, et al. [2013] for studies that looked at the short- and long-tern effects of this intervention). These studies established strong short-term effect which persisted when the subjects of the experiment were in their forties. In North Carolina an even more intensive intervention aimed at helping disadvantaged children, known as the Abecedarian Project, was introduced and evaluated (see for instance Campbell and Ramey 1994). Again, rigorous evaluations established positive impacts. These interventions to a large extent inspired the design of pre-school education, such as the Head Start Program, which is now very widespread. A bit later, a large project, known as the Nurse Family Partnership was deployed and evaluated extensively (see Olds et al. 2019).

And these types of interventions were not happening only in developed countries. In the 1970s some interesting studies conducted in Bogota and Cali, Colombia, showed, within proper randomized controlled trials, that targeted interventions, aimed at stimulating young children during the first few years of life, had considerable effects [see, for instance, McKay et al. (1978) for the Cali experiment and Waber et al. (1981) and Super et al. (1990) for the Bogota one].

In the late 1970s an influential study was developed in Jamaica, giving rise to what is now known as the Reach Up and Learn (RULe) intervention. As with the other interventions we mentioned, RULe was evaluated rigorously. Some of the results were presented in Grantham-McGregor et al. (1991). What is remarkable about the Jamaica study, which is now being replicated in a variety of contexts, is the fact the study children were followed for decades. As shown in Gertler et al. (2014), the impact of the intervention on a variety of outcomes, were very visible over 20 years after its end. And these types of interventions are replicated in a number of different contexts, from India (see Grantham McGregor et al. 2020; Meghir et al. 2023) and China (see Sylvia et al. 2020) to parts of Africa, and Latin America (Attanasio et al. 2022).

Given this evidence and much else, there is now a widespread consensus about the importance of the early years. Not only is it recognized that they have long-term effects; it is also widely accepted that they are malleable, so that well-designed intervention can change individual lives in the long term. There are, however, a number of big challenges to the goal of using these policies to guarantee that destiny is not determined by birth and that all individuals have a chance to develop fully. Given these challenges, the ghost of a fractured society made of different classes of individuals permanently divided even across generations is clear and present.

Words of Caution and Challenges

From a policy perspective, a big challenge is the implementation at scale of the early years interventions that I mentioned above. In addition to the financial costs that such interventions imply, there are several other challenges. First, while much progress has been made on the understanding of child development, there are still many unanswered questions about the details of the dynamics of the developmental process, where different skills interact at different ages to generate the outcomes of interest. A better understanding of this process can allow to target specific dimensions of child development and design more effective policies.

Second, as parenting have an important role in determining early year development, effective intervention should aim to change individual behavior, which might not be easy. As argued by Annette Lareau (2003) and more recently by Robert D. Putnam (2015) parenting, practices are driven not only by financial resources and individual preferences, but also by parental beliefs about the process of child development. If some parents do not "invest" in their children and do not stimulate them, it might not be that it is because they lack financial resources or do not love them. It could be because they might not think that certain activities are particularly useful. Individual perceptions might be different for parents with different backgrounds: Lareau calls the parenting practices followed by middle-class and rich parents "concerted cultivation," while poor and working-class families pursue a model of "natural growth," where children do not need special attentions and stimulation. Such different attitudes and strategies become even more polarized and distant when community become increasingly segregated and polarized.

Changing engrained beliefs is difficult and involves building interventions that can transmit the right messages appropriately. Changing behaviors might be even harder. Adults (rightly) do not like being lectured by individuals different from their immediate environment. It is necessary to identify agents in the community that are well-known and trusted.

Third, the availability of qualified (or qualifiable) labor force that can deliver this type of interventions is essential. From a policy point of view, it becomes important to provide the right set of incentives while at the same time it is necessary to provide the right set of qualifications.

An important implication of these considerations is that for the implementation of effective interventions at scale and in a sustainable fashion, it is necessary to rely on existing networks and infrastructures that are well integrated in local realities. Such challenges are behind the fact that certain interventions have been shown to be effective on a small and very controlled scale but turn out to be difficult to be deployed at scale. It is important to understand that, when failures are observed, they do not depend on the goal at hand being impossible to achieve, but on the way specific interventions were implemented and delivered.

Is This Relevant only for Early Childhood Interventions?

The challenges mentioned above are not only relevant for early childhood interventions, although, these should have central attention. These issues are relevant also for the design and implementation at scale of effective policies aimed at achieving a less fractured and polarized society. The way our societies have changed, partly because of the way technology has evolved and partly because of the way markets (local and international) are organized, has implied that many sectors of society have been left behind and excluded from the level of progress and growth experienced by other sectors. And it is not just a matter of material well-being: in many contexts entire neighborhoods or town have been left behind and the individuals that remained in them have lost a sense of dignity and belonging to a *society*.

The solution is not a regression to *the good old days*. There is no way to stop economic incentives and progress, and it is probably not desirable to do so: one should remember that while inequality might have increased *within* many countries, inequality across countries has declined remarkably and a large number of individuals have exited poverty. The challenge then is to redesign social and welfare policies not only to guarantee that certain needs are satisfied for everybody, but also to guarantee that society continues to work as such and that all individuals within a society get recognized as respectable and productive members of it. Such policies, which some have labeled as leading to *inclusive growth,* could avoid the extreme fragmentation we see in many places and, importantly, could increase social mobility. The role of communities should be rebuilt, so that the destiny of an individual is not determined by the place where he or she is born.

References

Almond, D. & Mazumder, B. (2013). "Fetal origins and parental responses". *The Annual Review of Economics*, 5(1): 37–56.

Attanasio, O., Bernal, R., Henningham, H., Meghir, C., Rubio Codina, M., & Pineda Ruiz, D.M. (2022). "Early stimulation and nutrition: The impacts of a scalable intervention". *Journal of the European Economic Association*, 20(4): 1395–1432.

Brookwood, M. (2021). *The Orphans of Davenport: Eugenics, the Great Depression, and the War over Children's Intelligence.* New York: Liveright.

Campbell, F.A. & Ramey, C.T. (1994). "Effects of early intervention on intellectual and academic achievement: A follow up study of children from low-income families". *Child Development*, 65(2): 684–698.

Gertler, P., Heckman J., Pinto, R., Zanolini, A., Vermeerch C., Walker, S., Chang, S., & Grantham-McGregor, S. (2014). "Labor market returns to an early childhood stimulation intervention in Jamaica". *Science*, 344(6187): 998–1001.

Grantham-McGregor, S., Adya, A., Attanasio, O., Augsburg, B., Behrman, J., Caeyers, B., Day, M., Jervis, P., Kochar, R., Makkar, P., Meghir, C., Phimister, A., Rubio Codina, M., & Vats, K. (2020). "Group delivery or home visits of early childhood stimulation: A cluster randomized control trial". *Pediatrics*, 146(6): e2020002725.

Grantham-McGregor, S., Powell, C., Walker, S., & Himes, J. (1991). "Nutritional supplementation, psychosocial stimulation, and mental development of stunted children: The Jamaican study". *Lancet*, 338(8758): 1–5.

Heckman, J., Moon, S.H., Pinto, R., Savelyev, P., & Yavitz, A. (2010). "Analyzing social experiments as implemented: A reexamination of the evidence from the High Scope Perry Preschool Program". *Quantitative Economics*, 1(1): 1–46.

Heckman, J., Pinto, R., & Savelyev, P. (2013). "Understanding the mechanisms through which an influential early childhood program boosted adult outcomes". *American Economic Review*, 103(6): 2052–2086.

Lareau, A. (2003). *Unequal Childhoods: Class, Race, and Family Life.* Berkeley, CA: University of California Press.

McKay, H., Sinisterra, L., McKay, A., Gomez, H. & Lloreda, P. (1978). "Improving cognitive ability in chronically deprived children". *Science*, 200(4339): 270–278.

Meghir, C., Attanasio, O., Jervis, P., Day, M., Makkar, P., Behrman, J., Gupta, P., Pal, R., Phimister, A., Vernekar, N., & Grantham-McGregor, S. (2023). "Early stimulation and enhanced preschool: A randomized trial". *Pediatrics*, 1(151) (Suppl 2): e2023060221H.

Olds, D., Kitzman, H., Anson, B., Smith, J., Knudtson, M., Miller, T., Cole, R., Hopfer, C., & Conti, G. (2019). "Prenatal and infancy nurse home visiting effects on mothers: 18-Year follow-up of a randomized trial". *Pediatrics*, 144(6): e20183889.

Putnam, R.D. (2015). *Our Kids: The American Dream in Crisis.* New York: Simon & Schuster.

Super, C.M., Herrera, M.G., & Mora, J.O. (1990). "Long-term effects of food supplementation and psychosocial intervention on the physical growth of Colombian infants at risk of malnutrition". *Child Development*, 61(1): 29–49.

Sylvia, S., Warrinnier, N., Renfu, L., Yue, A., Attanasio, O., Medina, A., & Rozelle, S. (2020). "From quantity to quality: Delivering a home-based parenting intervention through China's family planning cadres". *The Economic Journal*, 131(635): 1365–1400.

Waber, D.P., Vuori-Christiansen, L., Ortiz, N., Clement J.R., Christiansen, N.E., Mora, J.O., Reed, R.B., & Herrera, M.G. (1981). "Nutritional supplementation, maternal education and cognitive development of infants at risk of malnutrition". *American Journal of Clinical Nutrition*, 34(Suppl 4): 807–813.

Chapter 9

The Food of Tomorrow

Attilio Giacosa

World Demography and Food Availability

Food has always represented an essential element in human life, both for its crucial role in ensuring physical survival and for its emotional, social and cultural implications. Over the centuries the way in which we feed ourselves and the way in which food is produced and prepared has changed radically. Even today there are a large number of people who are not guaranteed a sufficient amount of food to feed themselves and this amount equalizes the number of obese people.

Therefore, the question to ask ourselves is: how can we make people more aware of the message "eat better," which for some means rather "eat less" and for others simply being able to eat?

How can we think about diet in terms of moderation and sobriety, without necessarily relating it to the manipulation of the body through sometimes dangerously extreme eating behaviors, for the frantic search for the look, the image-spectacle of the body?

Balanced diets, high-performance foods, synthetic additives and protein powders today guarantee the perfect functioning of a body-machine, which is increasingly in need of revisions and corrections. Daily renunciation, the intake of food simulacra (think of the so-called bars or sachets containing the most varied substances ranging from Baltic algae to freeze-dried vegetables) promise to keep you slim and fit, in reason for an interminable health battle to achieve an aesthetically correct, smooth, perfect body, but separated from the overall psycho-physical well-being. The exploit of attention and care towards foods increasingly designed according to the shape of the body makes us archive the desire for naturalness and curiosity, as well as wonder that food recalls and suggests, referring to primordial experiences of satisfaction and development, such as the rendezvous of the infant's mouth with the mother's nipple.

Given all this, it is estimated that the world population will reach ten billion people in 2050, with an annual increase in food requirements equal to 14 percent, and therefore an increase of more than double compared to the current one, by half of this century.

DOI: 10.4324/9781003498605-9

Will achieving this result be possible or is it a utopia? Is it a source of serene stimuli or catastrophism? Not only that, but the need to produce food useful for human health and at the same time produce food that guarantees the health of the planet, today threatened by profound climate changes and multiple environmental criticalities induced by human behavior, is fundamental.

One of the crucial aspects is linked to the world of agriculture. What appears to be a priority is not simply producing more but producing better and with more efficient agri-food systems than the current ones. For many years, various research centers and international organizations such as the FAO (Food and Agriculture Organization) and other United Nations units have been reporting that food production processes have a very critical environmental impact. Almost half of the planet's cultivable surface is now home to agricultural or livestock activities, with a withdrawal of fresh water equal to 70 percent of its global consumption and with an emission of greenhouse gases linked to the food chain which corresponds to 26 percent of the globality produced on our planet. In particular, it was observed that foods of animal origin and especially beef and sheep meat are in the top places for all indicators of environmental criticality, i.e., CO_2 emissions, use of land surface, energy consumption, potential for acidification of the environment and water eutrophication, or degeneration of water quality.

Let's hold on tight: in 2050, to guarantee nourishment for the entire planet, it will probably be normal to put on the table a dish of crickets or locusts or meat prepared in a laboratory or products printed with 3D technology, rather than steaks coming from cattle farms! Certainly, the transition to food choices that are much richer in products of plant origin is an obligatory path for most scientists who deal with agriculture, economics and health. This without necessarily becoming vegetarians in the strict sense or vegan. We therefore need to rethink food choices, we need innovation and research on the safety, effectiveness and cost of "novel foods" and the diet of tomorrow. This is what we know today and what we should and will do tomorrow.

"Zero Hunger"

The Sustainable Development Goal 2, known as "Zero Hunger", is one of the 17 sustainable development projects developed by the United Nations in the 2012 conference in Rio de Janeiro[1]. Subsequent pandemic events (primarily Covid-19 infection) wars and conflicts, climate change and an increase in social inequalities have further worsened the situation. Already in 2022 some 735 million people, corresponding to 9.2 percent of the world population, suffered from chronic hunger, and 2.4 billion people lived in conditions of more or less severe food insecurity. In addition to malnutrition, food deficiency leads to greater susceptibility to diseases, lower productive capacity and lower ability to improve one's life. In 2022 an estimated 148 million children were experiencing growth difficulties and 45 million children under the age of five had clear evidence of severe malnutrition. A world with "zero hunger" can have a positive impact on the economy, health, education, social development, socioeconomic equity and in particular gender equality; but this objective, to be achieved, requires a multi-dimensional approach with a profound review and transformation of global food systems, from the production of agricultural goods to consumption. For example, it would be useful, especially in the most fragile countries, to introduce crops resistant to water stress and also study crops resistant to heat stress.

The Various Players

The "alimentary person" is at the center of a system that sees the involvement of body, food, emotions, sensoriality, environment, culture, traditions, needs, and perspectives. This broad scenario is conditioned by various factors, which, like the players of a football team, participate with a very specific role, as if they had a number on their shirt and a position on the pitch established by an invisible coach. These players are represented by the lifespan which appears to be clearly increasing throughout the world; by the globalization of production systems and know-how; by information, by the media and "social media," with all the pros and cons that this aspect entails; by the profound climate changes and by the progressive deterioration of the health of the planet. Other players are the evolution of scientific knowledge in the medical field, such as the epidemiological and biological correlations between nutrition and health, nutrigenetics, nutrigenomics, the intestinal microbiota, junk food on the one hand and optimal diets – such as the Mediterranean diet – on the other, obesity, diabetes, tumors, gastrointestinal disorders of various kinds, cognitive decline and cardiovascular diseases. The same goes for the evolution of agricultural and industrial technologies, such as biotechnology which has opened up the development of genetically modified foods (GMOs), sparking wide social debate, owing to the birth of potentially great advantages and at the same time fears, with significant ethical and economic implications. If on the one hand the genetic manipulation of plants offers new advantages and opportunities in terms of production, with greater resistance to drought and diseases and reduction in the consumption of pesticides, on the other hand it is favoring the reduction of global plant biodiversity and is creating significant economic problems. This is due especially to the concentration of seed patents (even those not linked to genetic manipulation) in a few corporate groups, with obligatory and annual purchase of seeds at a high price, and with progressive indebtedness and impoverishment of vast areas of the planet usually already critical at the origin.

But the play maker, number 10, of this elusive football team, is agriculture.

A study by the United Nations Department of Economics and Social Affairs signed by Daniele Giovannucci and collaborators focuses attention on the contradictions of current agricultural practice. The study highlights how a high percentage of food produced in developing countries does not reach the distribution market, while in rich countries every year a share of food equal to the entire production of sub-Saharan African countries is wasted.

Every year, 20,000 to 50,000 sq km of productive land are lost as a result of soil erosion and degradation processes, especially in developing countries. Furthermore, with more than 50,000 potentially edible plants available on our planet, more than half of the world's agricultural production comes from four foods: sugar cane, corn, wheat and rice (Giovannucci et al. 2012).

The agricultural world on a global scale today sees incredible strategic and operational inequalities. Alongside regions in which the plow is still pulled by an ox, there are agricultural operations guided by AI, i.e., artificial intelligence, aimed at massive savings in fertilizers, pesticides and water, with reduction of management costs and less contamination of underground aquifers (Zha 2020). As already mentioned, agriculture is responsible for a significant impact on the environment, ranging from high water consumption to the marked production of greenhouse gases. Sustainable nutrition must refer to agricultural practices aimed at reducing its environmental impact. Precision agriculture already today allows us to monitor crops with drones and satellite technologies useful for promoting the yield and quality of the product with fewer resources, less consumption of pesticides and fertilizers and lower costs.

Vertical agriculture allows the cultivation of products indoors, with the use of multi-storey buildings, and a marked saving of agricultural surface area and an increase in the quantity of product per square meter of land surface. Other very innovative experiences come, for example, from the aquaponic technique and from meat or fish produced in the laboratory, as an alternative to animal breeding or traditional fish farming (Goddek et al. 2015). We will talk about these opportunities later.

Furthermore, these latter strategies are in line with the increasingly widespread aversion in public opinion towards intensive animal farming and animal slaughter practices. At the same time, the choice of foods of plant origin, such as vegetables, fruit, whole grains and legumes, compared to animal products with particular reference to meat and fish is also growing. This evolution of collective thinking will necessarily orient the agriculture of the future towards operational choices that are profoundly different from the current ones.

Sobriety is today the watchword for a more sustainable and quality diet. Many people, for example, say they want to eat less animal proteins and prefer products that are better for the environment and for taste. But wanting is not always doing! Consumers want to eat healthily, but also enjoy eating and, if possible, spend less.

Accustomed to eating everything, even what is not in season and at low cost, food quality is not always the main concern of the food industry and nutritional value is not necessarily a decisive criterion of choice for all consumers. But to "eat better," we need to know what we put on our plate.

Today, food supply is largely based on a competitive logic dominated by supermarkets, whose main argument is "the defense of consumers' purchasing power". This has involved the transition from a diet that comes from the land and offers great diversity thanks to the extreme variety of races and species, to an abundant, "standardized" and industrially produced diet with processed, or often ultra-processed foods and with potential increased risks to our health. Easy access to junk food, of which French fries are a typical emblem, and to high- or low-calorie snacks and supplementary bars or meal replacements and ready-to-eat products (such as soups or ready-made canned spaghetti) or instant cookables and sweetened and carbonated drinks can cause various health problems. Typical is the intake of foods with a high caloric density, that is, very rich in sugar and fat and poor in fiber and micronutrients. In this vast range of products there is often a high presence of colorants, food additives, thickeners, sweeteners, stabilizers, emulsifiers, flavorings, flavor and aroma enhancers, and trans fatty acids which very often play an "aesthetic" role more than nutritional and with an unfortunate impact on health. These developments have changed our relationship with food and profoundly modified the agro-industrial process and the role of agriculture in society.

The Big Five

There are five, or rather four plus one, food realities that could now favor greater food availability at low cost and with high nutritional value. The first four are jackfruit, algae, mycoproteins and jellyfish; the fifth is represented by insects.

The jackfruit (Artocarpus heterophyllus), also called Giaca or Català, originates in India, but is easily found throughout South-East Asia. This product is a synapse, containing multiple fruits that can weigh up to 30 kilos or more. It grows in warm climates and is easy to grow and very rich in nutrients, especially vitamin A, C, folic acid, B complex vitamins, minerals and especially potassium, magnesium, calcium, and iron. It has a sweet, fruity flavor, but its richness in fiber makes it a food with a low glycemic index and therefore also useful for those who are diabetic or at risk of developing this disease. The cooked seeds are very rich in proteins. Algae are already widely used in oriental cuisine, they are easily cultivable and rich in nutrients. Spirulina algae has even been identified by NASA for cultivation during long-term space missions. Mycoproteins are protein compounds derived from fungal, yeast and mold cells. A 500 kg ox produces approximately half a kilo of protein per day, while 500 kg of fungal cells produce

1,250 kg of protein in a day, with a cost 12,500 times lower. Jellyfish are easy to breed and destined for explosive spontaneous growth as a result of climate change. Their flavor is similar to that of seafood, such as oysters, and they are very rich in proteins and omega-3 fatty acids, with a low-calorie intake. Jellyfish should therefore no longer be seen as a critical source of trouble for swimmers, but as useful foods for diets rich in low-cost proteins and for low-calorie and slimming diets!

We will say a lot about insects in detail shortly, but it should be underlined that according to the FAO, two billion people already habitually use insects for food purposes. There are more than 1900 edible species. In 36 African countries, 527 different species of insects are consumed and the same happens in 23 countries in the Americas, and in 29 Asian countries[2]. Insects are a source of proteins of high nutritional value, comparable to those of meat and fish, with a flavor that often brings us back to well-known foods: the flavor of shrimp for crickets and of nuts for mealworms. Another element in their favor is eco-sustainability since they mainly feed on waste, such as rotting food.

Cultivate the Future

It will take a long time for its diffusion, but the future of food has taken an almost inevitable path that will lead us sooner or later to the availability and consumption of food produced in laboratories. Synthetic foods therefore, perhaps made with a 3D printer placed on the countertop of our kitchen next to the microwave oven and personalized according to the health needs of the individual. Alongside this, we will be able to find meat developed in the laboratory in the refrigerator at home, starting from animal stem cells. Not only or no more breeding of cattle or other animals, but meat made in test tubes and in quantities as desired. It will be sufficient to multiply stem cells in a culture environment in the presence of specific nutrients, without the need for antibiotics or other potentially critical substances for health and above all without wasting soil and without slaughtering practices (Choi et al. 2021). With a single starting cell, approximately 10,000 kg of fresh meat can be produced in a few weeks. The appearance, color, smell, and flavor are that of minced meat. In Singapore and the United States this product is already marketed. In Europe its commercialization is prohibited, but huge sums of money have been invested in research.

On the other hand, only half of the arable land is used for the production of food for direct human consumption. The remaining cultivated lands are used for animal feed. In this situation, the yield in the form of food consumed by humans is at most 20 percent of what the same surface would have produced in directly edible food, such as potatoes, courgettes, carrots, legumes, wheat, or rice. Similar projects are underway for the production of fish meat derived from the multiplication of fish stem cells. And not only!

The production of milk and dairy products produced in laboratories is now a reality. The environmental impact of dairy foods is not equal to that of meat, but still appears to be significant. In fact, to produce 1 kg of milk, an average of 628 liters of water and 9 sq meters of land are needed.

If there is still a long way to go for meat, fish and milk produced in a lab in terms of food safety and legislation, the potential use of insects as a food source for humans is very different. In fact, insect farming represents an ecologically valid and sustainable method for providing the world's population with a protein-rich food. This choice constitutes a low-cost method to reduce the shortage of protein products that afflicts large areas of the planet and at the same time create an alternative to Western diets very rich in meat. In 2021 the European Food Safety Authority authorized the marketing for human consumption of products based on larvae of Tenebrio molitor (yellow mealworm), Locusta migratoria (frozen, freeze-dried, in paste, or in powder) and larvae of Alphitobius diaperinus (mealworm). It is now also possible to sell cricket flour (Acheta domesticus) and all products containing it, such as sweets, chocolate, baked goods, sauces, meat substitutes and drinks, in all countries of the European community (Turck et al. 2021a, 2021b). The acceptance of these foods will not be easy because food is also culture and tradition, and this choice is currently foreign to the habits of most of the inhabitants of the planet. Furthermore, rigorous legislation is needed on production methods and accurate studies are required on medical safety and allergenic potential of the protein material of which these flours are made. It is a fact, however, that their diffusion will be increasing, both for animal and human consumption. So crickets for dinner, sooner or later!

Food Choices: What Is the Ideal Diet?

Food choices are strongly connected to a myriad of factors such as tradition, culture, socioeconomic status, health and psyche. It is therefore impossible to imagine an optimal diet that is the same for everyone. In 2019 the EAT-Lancet commission proposed a "flexitarian" diet, as an ideal choice for human and planetary health, based almost exclusively on plant foods with the occasional inclusion of meat, fish, and poultry (Willett et al. 2019; Rockstrom et al. 2023).

However, the possibility of universal application of this dietary model has given rise to many doubts in the international scientific community. In fact, we cannot ignore the various geographical and cultural contexts. While considering the need to increase the consumption of plant foods, for the future it seems logical to think, for example, of a wise integration of food models with significant use of products rich in proteins derived from insects in countries with low socioeconomic level and prevalent use of synthetic meat in Western countries, where the consumption of meat historically represents a habitual reality.

The ideal diet does not exist, but scientific research data have highlighted how it is preferable for human health to choose a diet rich in vegetables, fruit, dried fruit, nuts, cereals (preferably wholemeal), and legumes as a significant protein source. A diet rich in these foods, especially if associated with reduced consumption of meat and fats of animal origin, favors the reduction of the risk of many diseases and especially cardiovascular diseases and cancer. These are the basis of the Mediterranean diet, together with olive oil as an essential source of cooking and seasoning fats and two glasses of wine a day for males and one for females (when there are no religious or medical limitations) (Giacosa et al. 2013; Guasch-Ferré and Willett 2021). Even soy and its derivatives such as tofu and soy milk, algae, brown rice, turmeric and green tea represent a healthy and sustainable choice, rich in nutrients, antioxidants and proteins and typical of the culture of eastern countries. The integration of these Mediterranean and Asian food models is often defined as the "Mediterrasian diet" and also has a positive impact on sustainability and health of the planet and should constitute a reference model for everyone. It is therefore necessary to avoid the mortification of the senses, curiosity, research and the weakening of time for the consumption of the ritual meal. The transition from fast food to walking food (food to be consumed quickly while walking) is significant in a condition where time is increasingly accelerated and contracted and where the discomfort of the way in which food is ingested prevails, standing or sitting on walls, with fingers or with plastic cutlery. The aberrations of fast food have also affected early childhood foods, which are all, from chamomile to semolina, "instant," therefore deprived of a preparation time that allows for some form of dreaming state that can favor gradual access to the sense of reality and the dimension of caring (Schinaia et al. 2003).

The Problems That Need to Be Addressed

Food sustainability is undoubtedly the first problem that needs to be considered.

The concept of sustainability is now on everyone's lips and when talking about nutrition this target appears to be of great interest. In fact, as already mentioned, food production is responsible for a negative impact on the environment through the production of almost a third of global greenhouse gases. However, the positive information is that the foods that are least critical for the environment (i.e., vegetables) are those that are most beneficial to human health. Consequently, what is good for humans is good for the planet, not only in terms of the production of greenhouse gases, but also for the lower consumption of precious natural resources – such as water and soil – and for the lower use of energy and the reduced need for transport. According to the FAO, sustainable eating behavior includes abundance in foods with low environmental impact, promotes biodiversity, is culturally acceptable and fairer from an economic point of view and is safe and sound from a health point of view[3].

These indications point towards a diet with a predominantly vegetal content, rich in whole grains, vegetables, legumes, fruit, dried fruit and nuts and with a reduced intake of products of animal origin and above all meat, fish, eggs and milk. For the FAO, the cornerstones of food sustainability primarily include the support of small and medium-sized agricultural enterprises, the definition of objectives in terms of human nutrition rather than a simple increase in production, the orientation of research towards ecological and health objectives, the encouragement of technological development with applications suitable for use in differentiated socio-economic environments (including those that are culturally and economically weaker and backward), the reduction of waste and the limitation of soil use and of agricultural products for energy generation.

Alongside these aspects, the "material footprint" must be considered[4]. This is a model for measuring resource consumption and refers to various factors. In addition to food and drink used per capita, this analysis tool considers the material used for the home, daily mobility, tourism, electricity, heating and accommodation. The average value of the material footprint of the European citizen is 31,000 kg, but to be considered sustainable this value should be reduced to 8,000 kg on average, while considering its variability depending on the needs and lifestyle of each individual.

Another very important problem is related to climate change which is profoundly affecting the planet. High temperatures, progressive melting of glaciers, progressive drought, forest fires, less availability of fresh water, floods, increase in water temperature and rise in sea and ocean levels drastically affect life with progressive desertification of the Earth and lower productivity in agriculture, fishing and livestock breeding. Reduced food availability and increased poverty, hunger, malnutrition and diseases can be the obvious consequences of this climate trend. Drought negatively affects crops and pastures and favors the development of infesting parasites. Rising temperatures put the survival of many animal species at risk.

According to the United Nations[5], species are being lost at a rate 1,000 times higher than what is usually observed. Desertification arises not only as a result of the severe climate change underway in this historical period, but also as a consequence of incongruous human activities. It is the result of the extreme vulnerability of ecosystems in progressively arid areas, owing to excessive exploitation and inappropriate use of land. Poverty, political instability, deforestation, overgrazing, and inadequate irrigation practices are all factors that have a negative impact on land productivity.

Floods are also a problem because they can ruin crops and large agricultural and urban areas, destroying homes and goods with a further increase in the state of poverty and a stimulus to the migration of entire populations. During the decade 2010–19 it is estimated that critical events resulting from climate variations were responsible for the migration of approximately 23.1

million people on average every year, leaving an even larger number of individuals in a state of poverty. The majority of refugees come from the most fragile countries from an economic and organizational point of view and consequently less capable of responding to the effects of climate change.

In 2018 Jem Bendell, professor of Sustainability at the University of Cumbria, in the UK, published a scientific paper dedicated to "deep adaptation" to climate change, at his own expense. By "deep adaptation," Bendell means the need to prepare ourselves, as a human race, for the probable collapse of current society or even the extinction of the human species, since global warming and extreme meteorological events extreme weather events will increasingly disrupt social, economic and political systems, leading us to catastrophe. Bendell proposes an agenda for deep adaptation divided into four points: 1) Resilience, which consists of answering the question "how do we maintain what we really want to maintain?"; 2) Relinquishment, which asks us: "what should we leave behind so as not to make things worse?"; 3) Restoration, which is equivalent to asking "what can we recover that can help us against the difficulties and tragedies that are coming?"; and 4) Reconciliation, that is, "with what and with who shall we make peace with, as we awaken to the mortality of our mutual species?" Bendell talks about a return to compassion, curiosity and respect. This is a very stimulating approach to face the delicate future that awaits us.

For centuries we have become accustomed to the unlimited use of earth's resources. Now the situation is changing very quickly and negatively. Accepting this historical moment and its difficulties is a first step, but accepting – in agreement with Bendell – does not mean that things cannot change in the future. Now we need a great individual and social cultural change and the international community must review the programmatic and political approach to the problems imposed by these new and growing environmental criticalities and it needs to be achieved in the short term.

Energy choices constitute another thorny problem. Agriculture still consumes a lot of fossil energy (Paris et al. 2022). Direct energy is almost entirely diesel, petrol, gas, heating oil or electricity, especially to power tractors and motor mowers and other tools and machinery of various functions. In the agricultural sector it is important to increasingly satisfy energy needs by using renewable energy. But, as has been said, it is necessary to progressively limit the use of soil and agricultural products such as corn to produce energy. It is necessary to enhance wind power plants and solar panel stands with constant and preferential use of farm roofs as installation sites or to use systems that allow the development of agricultural crops underneath the plants. There is also a need for marked technological development of agricultural tools and machinery, including tractors, which use electric batteries and can operate using energy obtained from renewable sources.

The fight against food waste is a chapter that requires targeted interventions to make the agri-food system sustainable and to optimize the use of resources, reducing costs. We have already said about the extent of the problem at a global level: just think that every year, according to data collected by Eurostat, about a third of the food produced is wasted[6]. What is certain is that to reach this target, multiple interventions dedicated to each step of the entire production process must be activated, involving both the agricultural phase and the transformation of the products, packaging, transport and final consumption. It is necessary to launch targeted research and innovation programs that address the needs of individual environmental, geographical, socioeconomic and cultural realities. The rural African village scorched by the sun, the Scandinavian city immersed in woods and snow-covered for most of the year, the European industrial city or the Chinese megalopolis, New York, Mexico City or Cairo present profoundly different characters and needs! In this sector, research and development must make important progress.

In nature nothing is lost: there is no waste. Everything is recycled. Inspired by this principle, many companies are developing innovative by-products with food processing waste, initially intended to be eliminated. One of these is the creation of protein bars and other products derived from cereal waste from beer production. Similar processes are being developed in various large catering chains, as well as in restaurants and bars, with a significant impact also on cost reduction.

However, when talking about the fight against food waste, it is not necessary to refer only to the agricultural and industrial world. Every single individual must be primarily involved. Here are some tips for reducing waste at home suggested by the Eat-Lancet Commission in 2019:

- Buy what you really need, you can help yourself by writing a shopping list and planning meals.
- At the supermarket, buy products that are closest to their expiry date, especially if you know you will consume them in the short term.
- Don't throw away leftovers from a meal, you can use them as ingredients for other recipes.
- In the pantry, place products with a close expiry date in front of those with a more distant date.
- Freeze any leftover food, you can do this remembering not to leave it out of the fridge for too long once cooked.
- Check the supermarket to see if there are any products on offer with a short expiry date.
- Remember that the wording on the label "best before" is not an expiry date. The product is still safe for health, but some organoleptic characteristics may be lost after that date.
- Arrange food correctly in the refrigerator.
- When you finish a product, pay attention to the type of packaging and follow the rules for correct separate waste collection. Buy products with recyclable packaging.

- Don't buy too much food, especially fresh, if you're not sure you'll consume it all.

A further way to reduce waste is to limit individual food consumption for those who use it in excess, primarily those who are overweight. This data is almost never reported, but – considering the presence of more than a billion obese or markedly overweight people on a global scale – it is logical to believe that the excessive calorie consumption of this mass of individuals represents real food waste, given their use in excess of needs. Educational programs aimed at promoting the containment of food intake in obese people, in addition to promoting their health, increases the availability of food for the community, reducing waste from improper consumption.

The behaviors mentioned above seem trivial, but – if adopted on a large scale – they can significantly promote the reduction of food waste, provided that only substantial modifications of the entire food chain will be able to guarantee optimal results.

The "Vision" of the Food Future

The cocktail that sees the relationship between the human being, his health, his right to food availability and his freedom of choice on the one hand and the health of the planet, the quality and quantity of the food produced, the agricultural development and the evolution of food technology on the other represents the great challenge of the coming decades and must not see us unprepared or inadequate. The word catastrophism is meaningless in this area because knowledge of the facts and the possibility of planning appropriate choices allow us to avoid the collapse of our food future. What is certain is that we need to speed up international consensus on food policies and grow the culture of consumers, farmers and industries in the sector to guarantee a fair and sustainable life for future generations.

But how will it be possible to feed seven – and soon ten – billion people? The food of the future will not be able to do without technological intervention, meat and fish will be made in the laboratory, obtaining meat without breeding and slaughter and a myriad of novel foods will appear on the world food scene, saving consumption of land and water and critical emissions. Ludwig Andreas Feuerbach stated in the 19th century that we are what we eat, but today we can add that the choice of what we eat can change the world. Sustainability, ethics and health must be the inspiring principles of trends and food innovation and production for generations to come.

In conclusion, I think we should return to a simple and natural way of cooking, which has the right timing, which uses fresh and local products, changing the conditions for eating better and responsibly and introducing food education into schools. Furthermore, the ritual of lunch, or rather more and more of dinner, owing to working hours, should be restored, giving space to communication, conversation and not be mortified by watching the news or by the paroxysmal use of the mobile phone during meals. which prevents emotional and affective contact between guests. At the same time, however, we must realize that the food of the future is destined to be very different from what we know today. Food innovations are promoting the development of more sustainable, safe and efficient food production, while new technologies are transforming the way we prepare and consume food. While there are challenges to face and broad social, political and economic debates to advance, the future of food offers myriad opportunities to improve our food experience and promote a healthier planet for all, nothing but catastrophism!

Notes

1   United Nations (2012). "The global goals of the United Nations: 2. Zero hunger". www.globalgoals.org/goals/2-zero-hunger.
2   FAO (2021). "Looking at edible insects from a food safety perspective. Challenges and opportunities for the sector". https://doi.org/10.4060/cb4094en.
3   FAO (2021). "Sustainable Food and Agriculture". www.fao.org/sustainability/en/?utm_source=twitter&utm_medium=social+media&utm_campaign=faodg.
4   Sustainable Development Goal Indicators website of the United Nations. https://unstats.un.org/sdgs/iaeg-sdgs/2025-comprehensive-review.
5   United Nations Convention to Combat Desertification (2022). www.unccd.int/land-and-life/desertification/overview.
6   Eurostat: Food waste per capita in the EU in 2021. ec.europa.eu/eurostat/web/products-eurostat-news/w/ddn-20230929.

References

Bendell, J. (2018). "Deep Adaptation: A Map for Navigating Climate Tragedy". *IFLAS Occasional Paper* 2. www.iflas.info July 27, 2018.
Choi, K.-H., Yoon, J., Kim, M., Lee, H.J., Jeong, J., Ryu, M., Jo, C., and Lee, C-K. (2021). "Muscle stem cell isolation and in vitro culture for meat production: A methodological review". *Comprehensive Reviews in Food Science and Food Safety*, 20(1): 429–457.
Giacosa, A., Barale, R., Bavaresco, L., Gatenby, P., Gerbi, V., Janssens, J., Johnston, B., Kas, K., La Vecchia, C., Mainguet, P., Morazzoni, P., Negri, E., Pelucchi, C., Pezzotti, M., and Rondanelli, M. (2013). "Cancer prevention in Europe: The Mediterranean diet as a protective choice". *European Journal of Cancer Prevention*, 22(1): 90–95. doi:10.1097/CEJ.0b013e328354d2d7.
Giovannucci, D., Scherr, S., Nierenberg, D., Hebebrand, C., Shapiro, J., Milder, J., and Wheeler, K. (2012). "Food and Agriculture: The future of sustainability". *United Nations Department of Economic and Social Affairs*. https://sustainabledevelopment.un.org/content/documents/agriculture_and_food_the_future_of_sustainability_web.pdf.
Goddek, S., Delaide, B., Mankasingh, U., Ragnarsdottir, K. V., Jijakli, H., and Thorarinsdottir, R. (2015). "Challenges of sustainable and commercial aquaponics". *Sustainability*, 7(4): 4199–4224.

Guasch-Ferré. M. and Willett, W. (2021). "The Mediterranean diet and health: A comprehensive overview". *Journal of Internal Medicine,290*(3): 549–566. doi:10.1111/joim.13333.

Paris, B., Vandorou, F., Balafoutis, A.T., Vaiopoulos, K., Kyriakarakos, G., Manolakos, D., and Papadakis, G. (2022). "Energy Use in Open-Field Agriculture in the EU: A Critical Review Recommending Energy Efficiency Measures and Renewable Energy Sources Adoption". *Renewable and Sustainable Energy Reviews*, 158: 112098.

Rockstrom, J., Thilsted, S., Willett, W., Gordon, L., et al. (2023). "EAT–Lancet Commission 2.0: securing a just transition to healthy, environmentally sustainable diets for all". *Lancet*, 402(10399): 353–354.

Schinaia, C., Ciliberti, P., and Ferroni, M.A. (2003). "Cibo mutante per nuovi profili corporei". *Il Vaso di Pandora*, 11(3): 31–38.

Turck, D., Bohn, T., Castenmiller, J., De Henauw, S., Ildico Hirsch-Ernst, K., Maciuk, A., and Mangelsdorf, I. (2021b). "Safety of frozen and dried formulations from whole house crickets (Acheta domesticus) as a Novel food pursuant to Regulation (EU) 2015/2283". EFSA Panel on Nutrition, Novel Foods and Food Allergens (NDA). doi:10.2903/j.efsa.2021.

Turck, D., Castenmiller, J., De Henauw, S., Ildico Hirsch-Ernst, K., Kearney. J., et al. (2021a). "Safety of dried yellow mealworm (Tenebrio molitor larva) as a novel food pursuant to Regulation (EU) 2015/2283". EFSA Panel on Nutrition, Novel Foods and Food Allergens (NDA). doi:10.2903/j.efsa.2021.6343.

Willett, W., Rockström, J., Loken, B., Springmann, M., Lang, T., Vermeulen, S., Garnett, T., Tilman, D., and DeClerck, F. (2019). "Food in the Anthropocene: The EAT-Lancet Commission on healthy diets from sustainable food systems". *Lancet*, 393(10170): 447–492. doi:10.1016/S0140-6736(18)31788-4.

Zha, J. (2020). "Artificial Intelligence in Agriculture". *Journal of Physics: Conference Series, 1693*(1). doi:10.1088/1742-6596/1693/1/012058.

Chapter 10

Pushing Back on Catastrophism: The Case for a New Nature Narrative

Mark Halle

The Role of Narrative in Shifting Values and Behavior

Across the span of human history regular shifts have occurred in the way we understand the world around us, and in how we frame and receive fundamental ideas about how society and the economy work or should work. How we read and understand the world around us largely determines how we approach problem solving, how we set priorities, and how we seek to bring about the changes we wish to engender.

Since the first half of the 20th century and through a swelling body of research and literature (Campbell 1949, 1990), we have increasingly come to understand not only the role of a shift in narrative in bringing about change, but how articulating, promoting, and establishing new narratives can serve as an essential precursor and contributor to such change. This is true whether the change is fundamental and deep-rooted, or more simply a change in social norms and in accepted behavior.

The aim of this chapter is to unpack the importance of narrative in the process of human change. Taking the case of the prevailing nature narrative, it argues that a narrative based on frightening people into action has not worked and never will. Instead, a new nature narrative is needed. It must mobilize and offer hope and pathways to positive action at every scale. A successful narrative must not simply identify how we should behave in order optimally to advance the well-being of humanity, but rather identify pathways that lead to shifting the way values are reflected in human practice, offering a new framework and set of tools for addressing the world around us, thereby contributing to the growth of agency in individuals and communities – the fundamental measure of development success.

In this context, "'narrative' means 'Big Story', rooted in shared values and common themes, that influences how audiences process information and make decisions."[1]

Establishing a winning new narrative is not a straightforward, linear process. The relationship between an emerging narrative and achieving the desired change is not a simple one of cause and effect. Narratives are living

DOI: 10.4324/9781003498605-10

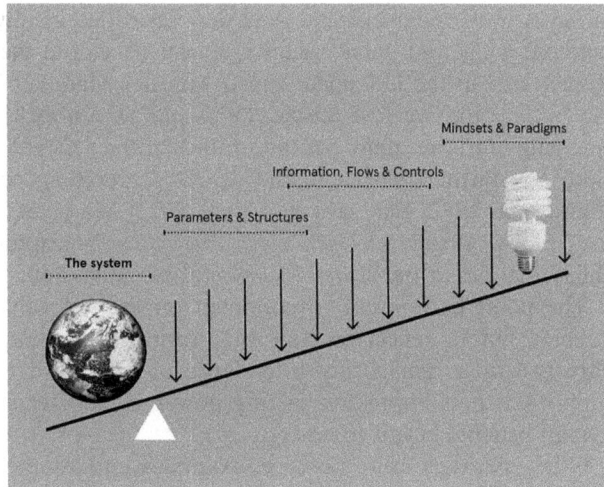

NARRATIVE SHIFT PRECEDES PARADIGM CHANGE.

Mindsets & Paradigms

Information, Flows & Controls

Parameters & Structures

The system

Figure 10.1 Narrative shift precedes paradigm change

and messy things, and their contribution to change is subject to many influences, from the dynamics of the information system to the enigmas of our evolutionary biology. However, the notion of "value shift" brought about by new narratives is solidly supported by the academic literature which, notably, analyses how intergenerational and inter-cultural breakthroughs take place.

Illustrative Examples of Narrative Shift

The examples that follow briefly illustrate how a dominant narrative can result in sometimes deep changes in human behavior, accepted social norms, or public policy. Each of these examples can be read and understood in different ways, and the purpose here is not to dissect each to examine its key features. It is, instead, to show that a changing narrative can result in serious change on a large scale.

To begin with an example of a global economic paradigm shift, it is interesting to reflect on the rapid emergence and dominance of economic arrangements centered on ensuring macro stability and allowing markets to set prices, rooted in what became known as the "Washington Consensus,"[2] and often more broadly described as "neoliberal economics." Coming on the back of seemingly endless economic chaos in the 1970s and 80s, this package of economic policies appeared to offer order, discipline, and the unleashing of the economic power of finance and investment in favor of those countries that chose to adopt it.

Some would argue that these economic policies are nothing more than a variant on ideas that had their origin with the Industrial Revolution, or even earlier in the Age of Enlightenment. And they benefitted greatly from the strong, corporate-friendly and public sector skeptical policies in the USA under President Reagan and in the UK under Prime Minister Margaret Thatcher. In the USA, the notion that we had reached "the end of history," with the world finally emerging onto the sunny plateau of unchallenged capitalism, was an attempt to close any further debate on the value of prevailing economic arrangements. (Fukuyama 1992; Bhargava and Luce 2023) Margaret Thatcher's wide use of the "TINA" slogan – "*There is No Alternative!*" – captured the sense of inevitability that lies at the heart of successful narrative shift.

The point here is not to comment on the net value or impacts of these policies, but to reflect on how they came to dominate the macroeconomic narrative in a remarkably brief lapse of time, and how the narrative has supported their impressive lasting power, even after strong doubts on their overall benefits began to emerge.

Other, perhaps more positive, examples underline the power of narrative shift. Slavery represents a barbaric chapter in human history; and yet it represented the chosen form of agricultural production in large parts of the world for a quarter of a millennium. Slavery was so entrenched – and so profitable – that its proponents warned of economic collapse if it were to end. And yet when it came, the end of slavery in the US, the UK, France, and other countries took place over what was, historically, a very short period. Many factors contributed to preparing the ground for the end of slavery, including the technological advances that characterized the Industrial Revolution. It also occurred because what had previously been considered an "externality" (the suffering of the slaves) gained weight and finally prevailed

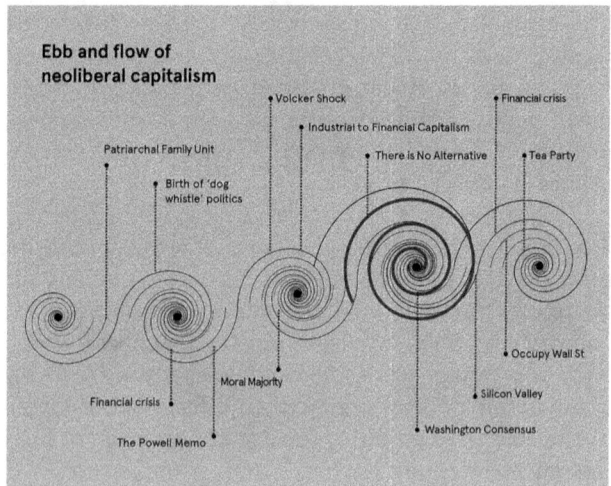

Figure 10.2 Adoption of a new narrative follows a messy evolution

over the economic case for slavery's continuance. One must not under-estimate, however, the growing power of the anti-Slavery narrative which began to take hold and bite, projecting a sense of inevitability that slavery would soon be phased out. The narrative contributed strongly to an acceleration of the events that led to this outcome.

Two other examples are drawn from the more recent past. Smoking tobacco in public was commonplace across the planet not many decades ago. Efforts to limit it were dismissed as entirely utopian such was the power of tobacco lobbies over public policy decision-making. And yet, in most parts of the world, the end of smoking in public came quickly and behavior chan-ged so fundamentally that violating smoking bans is today widely regarded as strongly anti-social – even shocking – behavior, even though smoking in public was the deeply established norm not long before.

Second, the shift away from lax enforcement of rules against driving under the influence of alcohol is a similarly powerful case, especially as it was not a change of regulation that principally drove the social change, but a narrative-based shift in expected and acceptable behavior.

Many factors converged to contribute to the tipping point being reached in these latter two cases. In the first, scientific evidence accumulated to the point where it was no longer credible to deny the link between smoking and negative health impacts. Emerging science on the impact of secondary smoke put paid to the notion that smoking was simply a life-style choice that con-cerned only the individual smokers themselves. Public opposition to smoking grew to the point where sound electoral politics dictated support to public smoking bans rather than bowing to tobacco lobbies. And legal action

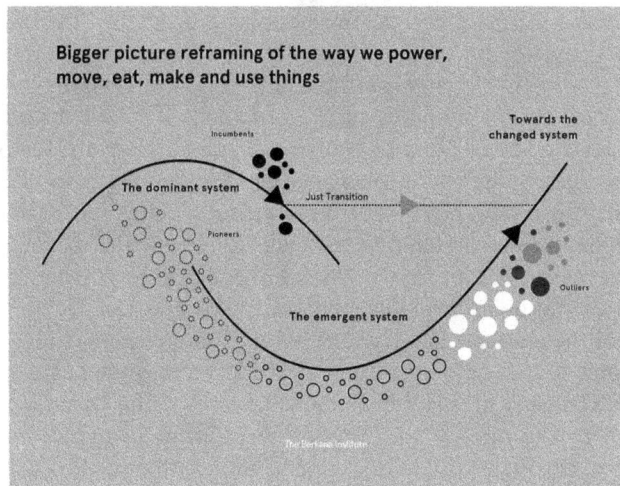

CAPTURES THE SENSE OF INEVITABILITY.

Figure 10.3 New narrative surges from below as the old one declines

against tobacco giants greatly undermined their reputation by bringing to light some seriously criminal behavior on their part.

In the case of drink driving, a mobilization led by the families of victims demanded attention to the issue and persisted until the legal norms were changed. Student drivers were confronted by graphic illustration of the consequences of driving under the influence. As with smoking, a new narrative – a new social norm – took over and is now solidly entrenched (in those countries that adopted it). The growing prevalence of this new norm preceded and contributed strongly to the legislation that codified it.

These few examples, picked from among an interminable list of candidates, indicate how those with story-telling ability can influence the way issues and options are framed, shape the discourse around them, and control how information is presented to the public. A well-articulated narrative arising from this process can replace established ones, leading to a deep change in the way that issues are understood, approached, and managed, as, for example, Jonah Sachs (2012) says.

Narrative Shift Mechanics

Building a new narrative – and navigating it to the point where it begins to prevail over earlier narratives, is both an art and a science, and a good deal of research has gone into the process. It is a science because much is now understood about the mechanics of narrative shift. One key insight, illustrated below in the graphic drawn from the Black Lives Matter movement (Centola 2021), is that key elements of an emerging narrative gather and grow at the periphery of an issue, sometimes establishing connections casually, sometimes deliberately. The narrative shift begins in earnest only as serious connections on the periphery begin to "tip inwards" to overwhelm and replace the notions that occupied the space previously.

Elements of a new narrative are, almost necessarily, obliged to take up residence on the periphery of the prevailing story set, because they question and challenge it but do not have the strength to overcome the power of the narrative in place. It is only when narrative elements begin to cluster and interact that they gain the power and confidence to emerge, federate and move to the tipping points in which they begin to replace the received wisdom of the older narrative. The tipping point for the Black Lives Matter campaign may have been the police brutality that led to the death of George Floyd in Minnesota, but the elements had been clustering and connecting at the edge of the issue for some time. The trigger event tipped them into the heart of the issue.

The next illustration demonstrates how the core ideas of a new narrative expand to occupy a far broader field in the imagination, the public discourse and, through professional attention, become the accepted norm.

As noted above, narrative shift is not merely a science – an established set of objective steps that plot a pathway from one narrative to a new one in a

TIPS INWARDS AS CLUSTERS CONNECT ON THE PERIPHERY.

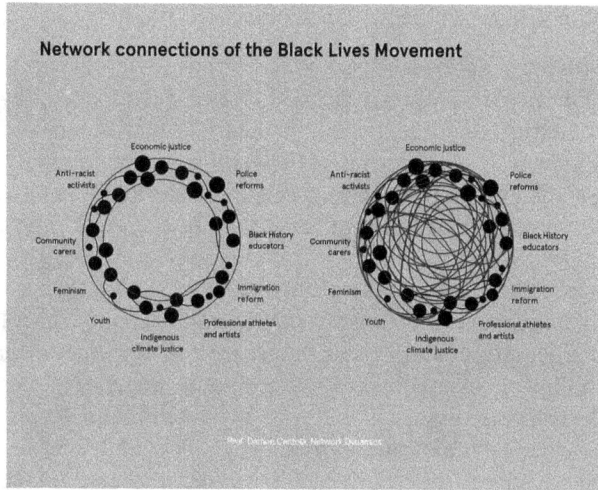

Figure 10.4 Connections at first on the periphery, then tip inward

NEW 'COMMON SENSE' THAT EXPANDS WHAT IS POSSIBLE.

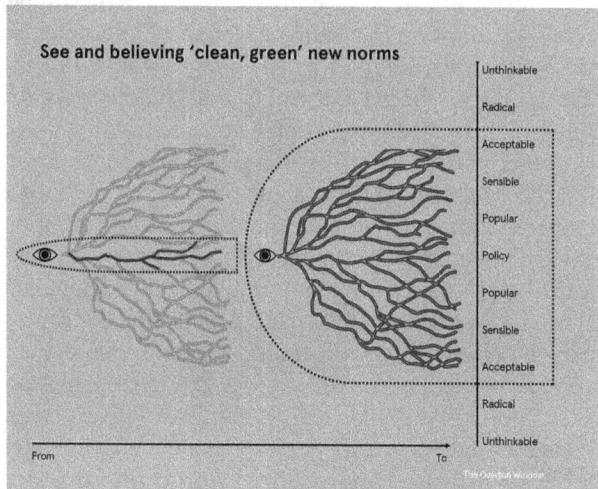

Figure 10.5 New narratives gain acceptance by an expanding sense of what is possible

linear and predictable fashion. Instead, a new narrative originates in a sense of dissatisfaction with established thinking, with values that do not or no longer find reflection in the prevailing narrative, or with a deft sniffing of the winds to detect coming changes. These nodes begin to connect and find common purpose, until they reach critical mass and deliver a range of "stronger stories" that set the tone and establish the content of the new narrative.

The Hope vs Fear Dilemma

Moving from specific examples of narrative change to meta-narratives that underpin the future of humanity, two narratives today dominate public discourse around the linked planetary threats of climate change and nature destruction. Both narratives are still strongly dominated by the invocation of increasingly precise science-based scenarios to paint a stark picture of the consequences of inaction (or inadequate action). The climate world paints a picture of rapid global warming, sea level rise, disruption of established weather systems, rapid growth in extreme weather events and, before long, the triggering of changes that can no longer be controlled or reversed.

The nature world relies on a narrative warning of mass extinction, impoverishment of entire ecosystems, and impending ecological collapse that threatens to disrupt global food production and access to essential ecosystem services such as water supply. At its best and least fear-based, the narrative demonstrates the beauty, complexity, and sheer marvel of nature (e.g., David Attenborough's stunning television series) and argue that it would be a pity to lose such extraordinary gifts.

The two narratives have in common that they appear to rest on a fundamental assumption that humans, faced with looming catastrophe, will rapidly act in their own long-term self-interest – at scale, and before it is too late. This overlooks the evident reality that there is always an alternative – that of denial. Far from driving people to positive and timely action, the result of fear-based narrative is often to generate a sense of hopelessness – the sense that the problems are too great for any individual to make a difference or that it is too late to act; faith must instead be placed in others with the means and the capacity to drive change at scale, or on as yet imprecise developments – such as miraculous technological developments that will engender a reversal of present trends. This denial is a dangerous transfer of responsibility to often unnamed others and undermines the power of narrative to mobilize individuals to bring about the change that is needed.

Beginning in the first half of the 20th century, it is now understood that successful narratives are built not only on the replacement of prevailing stories with fresher and stronger stories that play to emerging memes and notions, but that they follow a series of well-established steps (often known as The Hero's Journey) leading to successful outcomes in which the hero finally prevails over adversity. The steps in the hero's journey provide the template for endless books, films, plays and other forms of art that involve overcoming looming threats and the triumph of good over evil.

A lesson that can be drawn from this is that successful narrative shift requires striking an optimal balance between articulating a credible threat (justifying the need to take action, sometimes urgently) and the opportunity – the desirable outcome that is not only within reach but whose achievement must move from being a possibility to becoming inexorable. The failure of the prevailing climate and nature narratives is that they have proved unable to generate a narrative of hope, a mobilizing narrative in which everyone wishes to play an active part.

The Weakness of the Current Prevailing Nature Narrative?

The focus of this section is on how to bring about a shift in the nature narrative. The climate challenge addresses invisibles (e.g. greenhouse gases), is chopped into siloed narratives around the energy transition or technological development and is mostly divorced from a strong sense of embedded values that help drive human behavior. Nature, on the other hand, appears to offer more promising opportunities for narrative shift for two reasons – first, it is strongly value-based, connecting with impulses hard-wired into humans by millennia of evolution; and second, it offers the most promising opportunities for rapid and positive change.

As noted above, the prevailing nature narrative relies on raising alarm at the impending catastrophe and its impact. It focuses on nature loss, the breakdown in ecosystem services, the depletion of economic resources such as ocean fisheries, the accelerating rate of extinction, and the threat of irreversible damage. Not only does this fail to generate action at scale, it is surprisingly mute in respect of the range of actions available to address the problem. The narrative is also inward turned, the dogma of a conservation priesthood that preaches in an "echo chamber". In the end, the repeated failures of nature conservation action offered as the remedy for the crisis that the narrative advances undermine trust in established narrators in government, civil society, or the media, who are regarded as increasingly unreliable.

The prevailing nature narrative misses the reality that should be at its base – namely that, given a chance, nature will in most cases bounce back and restore itself. Unlike climate where excess carbon will remain in the atmosphere for decades after we cease to emit it, nature's self-repair begins almost immediately and can, if allowed, restore degraded land and ecosystems both to productivity and to biodiversity.

Elements of a New Nature Narrative

Nature is, potentially, a good news story but nobody attuned to the present narrative stream would ever guess it. The action needed to address nature loss is both positive and mobilizing, inspiring hope and replete with stories of regeneration. Further healthy nature, whether untouched or restored, provides the positive foundation on which to address both equity and climate action, leading to real expansion of human agency and, therefore, to human development.

What, then, might constitute some of the elements of a new nature narrative? It must, of course, start from the context of our current predicament and the urgency of taking action to counter present trends. However, this can only be the canvas on which the new narrative is painted and not, as at present, the painting itself, the central message. That message must plot a path to the deep currents that link humans with nature and its cycles. These connections – too often lost in the frantic immediacy of our daily lives and its distracting stimuli – provide a solid foundation of basic human values that lies at the root of all cultures and faith traditions, e.g. the principle that we must do no harm to others or to the earth around us (Wilson 1984; Shrikanth 2023).

But like life itself, it must also be rooted in the concept of rebirth and renewal. Nature regenerates if allowed to; it forgives; it offers a second chance; it is kind – offering shelter, sustenance, companionship. The new narrative must build not on the rearguard action of traditional conservation, trying to fight the losing battle to save the last remaining fragments of untouched landscapes or ecosystems. Instead, it must harness the new interest in regeneration, in ecological restoration, in rewilding – all focused on taking a bad situation and improving it (Tree 2019).

It must emphasize the opportunity to act at every level. Biodiversity can be restored in a home garden, on a balcony or in a city park. It can be part of a community or district plan or adopted at the level of a sub-national jurisdiction such as a state or province. Effective action can be taken by every one of us, in our neighborhoods, in our communities, in our churches, clubs or professional networks. It can be taken in cities, districts, countries, or we can act globally. There is a positive role for action and actors at every level of the pyramid.

The potential for national action is almost infinite. All this leads, like tributaries joining to form a powerful river, to generating an inexorable demand for a new relationship with the planet.

This must find strong reflection – through the manifestation of citizen and consumer demand – in the activities of corporations, banks and investors. The outlines of a "Nature Positive" standard – where the impact of all corporate or financial activity leaves nature better off than before that activity took place – are beginning to crystalize and can lead to a new norm being adopted on a broad scale. It is not difficult to imagine a world in which a corporation that destroys nature in pursuit of profit is reviled as, today, we would revile a company that employed slaves or children. Building the resilience of nature must become the norm in financial and corporate activity rather than the exception as we move to a world where nature is rebuilt in the normal course of economic activity.

This will happen only through an avalanche of positive and inspiring stories that highlight the world that is, at first, possible and, subsequently, inevitable. There are already thousands to draw from.

In summary, the purpose of a new narrative around nature-rooted development must be to provide a means for values to find better reflection in the behavior and choices made by individuals and communities.

Conclusion

With criticism of the prevailing neo-liberal economic model everywhere on the increase, and with its consequences on the planet and on human well-being increasingly a source of preoccupation, there is an opportunity now to target a major value shift, supported by a narrative that places planetary health and equity at the center of development purpose.

The world is ripe for a narrative shift that reconnects with the foundation of values that lead humans to seek and care for a vital connection with nature. However, this new narrative – badly needed to replace the power and dominance of neo-liberal thinking – cannot emerge if it continues to be based on the catastrophe-based narratives that dominate the world of climate and biodiversity.

A successful new narrative must build on the notion of hope and possibility (Solnit 2004). It must have the effect of mobilizing all of us by showing positive pathways forward, a building of momentum in this direction, and the emergence of communities of practice ready to roll up their sleeves and place the interests of a healthy planet front and center. Only a new narrative can ensure action at pace and scale.

Notes

1 https://belonging.berkeley.edu/shifting-narrative.
2 https://en.wikipedia.org/wiki/Washington_Consensus.

References

Bhargava, D. & Luce, S. (2023). *Practical Radicals: Seven Strategies to Change the World*. New York: The New Press.

Campbell, J. (1949). *The Hero with A Thousand Faces*. New York: Pantheon Books.

Campbell, J. (1990). *The Hero's Journey: Joseph Campbell on His Life and Work*. New York: HarperCollins.

Centola, D. (2021). *Change: How to Make Big Things Happen*. Boston, MA: Little, Brown Spark.

Jenkins, A. (2018). "Shifting the narrative". *Othering & Belonging Institute*. https://belonging.berkeley.edu/shifting-narrative.

Sachs, J. (2012). *Winning the Story Wars: Why Those Who Tell -and Live- the Best Stories Will Rule the Future*. Brighton, MA: Harvard Business Publishing.

Shrikanth, S. (2023). *The Case for Nature: Pioneering Solutions for a Planetary Crisis*. London: Duckworth.

Solnit, R. (2004). *Hope In the Dark: Untold Histories, Wild Possibilities*. London: Faber & Faber, 2016.

Tree, I. (2019). *Wilding: The Return of Nature to a British Farm*. London: Pan Macmillan.

Wilson, E.O. (1984). *Biophilia: The Human Bond with Other Species*. Brighton, MA: Harvard University Press.

Chapter 11

Catastrophism and Media Catastrophic Images

Cosimo Schinaia

Changes in Environmental Landscape

At one time the visual memory of an individual was limited to the heritage of his direct experiences and to a restricted repertory of images reflected in culture. The possibility of giving form to personal myths arose form the way in which the fragments of this memory came together in unexpected and evocative combinations (Calvino, 1988).

In 1976 Didier Anzieu described how the Eskimo community in the far north took advantage of the long winter period without sunshine or the prospect of hunting expeditions to dream together about how community life would resume in the summer; what marriages might be arranged; who would become the new head of the village community. The inhabitants of the igloo would tell each other their dreams in the morning after a night of bitter cold. They strung the dreams one after the other like pearls on a necklace, like a collective story of their social group. Their day would start well when someone, while waiting for the communal breakfast and the water to boil the dried herbs and frozen bacon slices, would tell how he dreamed about what he had done while awake, the conversations he had overheard, the tasks he had done. Such a dream received full attention if it told the dreamer and his company about the experiences of the previous day. This showed that these activities were not useless because, once they had finished, they were dreamed again. It was a way of feeling solidarity, and of taking a step back during the long hibernation.

From a psychoanalytic point of view, external reality is never a given, but rather something that continually acquires shape and meaning through its constant rediscovery and reorganization in internal space (Schinaia 2023).

Thus, the chain of dreams evoked in Anzieu's text could have the sense of maintaining the link with the reality of everyday life while living in a dormant and immobile nature (Franckx 2023). According to Anzieu, dreams referring to natural phenomena, such as the return of spring and the thaw, frozen rivers, dense forests, seal and whale watching, the midnight sun or the northern lights would have been of little interest because they concerned a

DOI: 10.4324/9781003498605-11

stable, reliable background. The Eskimos knew – Anzieu says – that they belonged to the cycles of the seasons, and they counted on the permanence of the cycle of nature.

This was the case in 1976, because today we are forced to think differently. Weather, once the basis of social, productive, and emotional stability, is turning into an emblem of uncertainty and disorientation. The unprecedented and accelerated scales of the change make it more of an "out of place" phenomenon than a basis for identification. People are now invaded by an environmental reality that is no longer constituted by the permanence and relative predictability of seasonal cycles, so our natural background is no longer immutable, a good and stable container, but kaleidoscopic and changing.

What was silent turns out to be active, vibrant, close rather than distant, foreign rather than familiar, powerful rather than under our mastery, and above all interrelated and interdependent, almost intimate, rather than outside society and culture (Van Aken 2020). The environment changes around us, betrays us, disorients us, uprooting us, in the most literal sense of the term leaving us physically without the earth in which to sink our roots (Revelli 2016).

Climate change is mainly affecting the Arctic regions, where, because of thermal inversion, glaciers are melting like snow in the sun, permafrost is disappearing at the speed of light, uncovering fossils millions of years old and organic matter from long-extinct plants, but also encouraging the release of large quantities of greenhouse gases (CO_2 and methane) into the atmosphere.

We can say that today the cycle of nature is impermanent. Nature that had been at one remove and considered as a "thing" infinitely available to us, now reappears with dynamically invasive characteristics. This disrupted and uncanny environment presents itself as "other" than us, in a continuous emergence of forms of threat. It follows that ecological catastrophes necessarily reorient psychoanalytic thinking about the psyche and its dynamics.

The Uncanny of Environmental Disastrous Images

Freud (1919) defines the "Uncanny" as what is frightening which goes back to what has been known for a long time, to what is familiar to us. He said that it is obvious to deduce that if something arouses fear it is precisely because it is known and familiar.

Unheimlich is the motion of the soul that we feel when we realize that there is no possibility of tracing our experience back to unambiguous terms and sharp, irreversible distinctions. When we discover that the very thing that seemed able to reassure us is the very thing that disturbs us. For example, plastic, overproduced, thrown away, and forgotten, which is a mundane part of our material culture for decades, reappears, re-emerges as uncanny after our expulsion and laborious disavowal.

The waste we had concealed and repressed reappears with new environmental relations, whereby the resulting interdependencies are constituted as disruptive threats. Climate change is implicitly disruptive: atmospheric conditions, and the carbon-intensive lifestyles that alter them are extremely familiar, and at the same time now present themselves as new threats and uncertainties (Ghosh 2016).

The external world phenomena (nature and landscapes for example) have always given expression and color to the mind's psychic states and inner fantasies. When confronted with nature (for example, a beautiful rainforest, or an arid desert), the psyche might generate mental images capable of communicating the incommunicable. Two painters, Edward Munch and Oswaldo Guayasamin have tried to represent the unrepresentable.

Expressionist painter Edvard Munch's iconic painting *The Scream* [1] is one of the symbols of the universal experience of the restlessness and despair. The distraught figure, with its contorted features and vivid colors, does not appear to be screaming but is widening its eyes, covering its ears, and opening its mouth in order to block out nature's powerful and heartbreaking scream. In Oswaldo Guayasamin's painting *El Grito,* [2] the subject screams in rage and pain, conveying the visceral suffering caused through violence and oppression. Fields of blue, grey, and yellow amplify the anxious mood and frame the subject's exaggerated mouth and impassioned scream to the world.

Clinical Experiences During Catastrophes

Research published in "Nature" by Helen Pearson (2024) shows how climate change affects mental health around the world – from lives that are disrupted by catastrophic weather to people who are anxious about the future. According to a global survey (Simon et al. 2022), the Philippines has the greatest number of young people experiencing high levels of climate anxiety, because the passage of Typhoon Haiyan in 2013 left many residents traumatized by the devastation. The first large-scale investigation of climate anxiety in children and young people globally (Hickman et al. 2021), however, documents how 75 percent of respondents across all countries, although they have not been subjected to extreme weather events, said that climate made hem think the future is frightening, and 56 percent said that it made them think that humanity is doomed. For all respondents, a perceived failure by governments to respond to the climate crisis is associated with increased distress.

In my clinical experience, when a disastrous flood occurred in Genoa in November 2011, many patients brought cataclysmic dreams and memories into the psychoanalytic session. In particular, one young patient, after reporting the fear he felt facing the risks of flooding while coming to the analyst's office, had a memory of an intensely traumatic event from his childhood.

When Alberto was a child, in the summer months he went on holiday to a seaside resort with his mother and two younger brothers. Dad used to visit them at weekends. His great and protective presence became an opportunity to play fearlessly together in the sea water. We children – he recalls – climbed on top of him, as if he were a big rock, and then dived into the water; then we would try to stay underwater in apnea as long as possible, and then emerge triumphant and ready to climb him again. Once, however, my father did not realize that we had strayed too far from the beach, so after a dive from his shoulders I no longer felt the sand under my feet. My father, who was caring for my younger brothers, had not noticed my difficulties. That water which until that moment had been the quiet witness of my joyful movements suddenly became dangerous. I started swallowing water and having trouble breathing. When my father realized that I was in danger of drowning, he picked me up and carried me to shore, helping me expel the water I had ingested. Cuddles and orange juice prepared by my mother allowed me to recover, smile and return to the water the next day.

The flood of those days in Genoa had caused the death by drowning of two children who were going to school accompanied by their parents, and the patient had been so affected by it that he remembered and relived the traumatic moments of the risk of his drowning. The presence of the father, perhaps initially careless, but then saving and protective, together with his mother's loving cuddles and orange juice refer to the containing experience of his analytic relationship which allowed him to work through his fear of being overwhelmed by his anxieties, as by the mass of water that flooded on the way to the session. The psychoanalyst can be the mediator, metaphorically the bridge, between internal reality and the environment, in an often painful, but necessary and fascinating cognitive adventure (Resnik 1990).

Another time, when the Morandi bridge[3] collapsed in Genoa in August 2018, causing 43 deaths, many patient's dream sequences depicted the collapse of the bridge as the very significant expression of their psychic breakdown, broken dreams, lost hopes, disappointed expectations, and failed projects. The bridge is the dream structure that indicates the path, the possibility of deepening an acquaintance, smoothing out divergent ideas, but above all the need to overcome one's limitations. The collapse of a bridge is an image that conveys a division, a separation, an interrupted communication, the unhinging of relationships hitherto felt to be stable.

In the aftermath of catastrophic events, the powerful images continuously spread by the media, the external reality becomes the iconic frame with a strong representative charge of confused, unthinkable, unrepresentable internal feelings. Women and men who did not have a direct relationship with the disaster, however, highlight in their psychoanalytic session experiences of psychic precariousness, of the cancellation of existential projects, of depression and anguish of death, of fear of the future, of the interruption of significant affective relationships, of irredeemable bereavements and losses. They can also experience resentful fantasies of vengeance, violent vindictive destructiveness and impotent need for justice against an invisible perpetrator whose power over them seems impossible to escape. Secondary traumatic stress can occur at different levels in people who have not experienced the disaster, because the constant media stream brings with it stress and uncertainty.

According to Winnicott (1974), clinical fear of breakdown is the fear of a breakdown that has already been experienced, a breakdown of the establishment of the unit self, related to environmental vagaries or failures. "The breakdown, a fear of which destroys his or her life, has already been. It is a fact that is carried round hidden away in the unconscious" (p. 104).

The possibility of representing internal traumatic vicissitudes through the structuring images (Correale 2024) of memories or dreams of catastrophic events recounted and dramatized in the psychoanalytic session allows for their initial working through thanks to the stability of the setting as facilitating environment, the containing capacity of the analyst's mind, and the creative meeting of the two minds, that of the analyst and of the patient. Through the acknowledgement of the intensity of internal reality, an initial process of mental alertness, of recognizing the gravity of external reality becomes possible (Schinaia 2019).

According to Susan Kassouf (2022), cataclysmic realities of climate change call upon all of us to cultivate catastrophic thinking and develop "traumatized sensibility," that may offer one way to counter the unthinking states. She suggests that sometimes people are able to learn from their experiences of trauma in ways that disrupt the culturally dominant an environmental orientation, that is, an orientation that brackets out the more-than-human environment.

Developmentally, it is crucial for the acquisition of a healthy understanding of reality, both internal and external, to have a mature enough object/mind to depend upon. Alterations of external reality concerning the permanence and impermanence of the cycle of nature bear directly on our internal world, and vice-versa. Human development is predicated on the need for a stable object. Without it, our humanity is endangered (Schinaia and Todes 2024).

The potential creation of a stable object helps alleviate overwhelming eco-grief that may otherwise result in severe solastalgia (Albrecht 2005, 2020)[4] that could be associated with an increased suicide risk (Radua et al. 2024).

The Invasive Power of Images

Modern consciousness and the perception of reality have gradually developed through the undisputed dominance of the sense of sight.

Since the times of classical Greece, thought has been based on vision: Plato considered sight to be the most important faculty of humans because it is closest to the intellect, and he separated sensation from thought. Similarly, Aristotle regarded sight as the noblest of the senses because it perceives the most immaterial aspects and is thus closer to intellect. During the Renaissance, then, the representation of space was codified with the invention of perspective, which made the eye the central point of the perceived world, thus associating it with self-cognition (Martellotti 2004).

In recent decades, however, the power of the visual is greater than ever before, so we witness such a surge in the number of visual images that our lives seem to be already saturated with them. From Google to Instagram, to video games, to art installations, this transformation is giving rise to a global visual society that generates confusion: it is liberating and disturbing at the same time, as we do not merely see the world, but rather keep reproducing it in images that we share and exchange with others. Our effort to understand the reality that surrounds us is primarily visual, and even our effort to modify it is part of our visual culture (Mirzoeff 2015).

In modern society, communication is channeled through mass media filter information. However, these filters often highlight or even invent destructiveness underlying events, fostering the spread of what is termed as "collective psychic infections" (Zoja 2011).

In recent years we have been pelted with warnings of global catastrophes, as overblown by the media as they are quick to disappear from memory. According to Calvino (1988), we live in an unending rainfall of images. The most powerful media transform the world into images and multiply it through the phantasmagoric play of mirrors. These are images stripped of the inner inevitability that ought to mark every image as form and as meaning, as claim on the attention and as a source of possible meanings. We are bombarded today by such a quantity of images that we can no longer distinguish direct experience from what *we* have seen for a few seconds on television, and the images we encounter in modern society lack the inner necessity that should characterize every image, both in terms of form and meaning. Much of this cloud of images fades at once like dreams that leave no trace in memory; but a feeling of alienation and discomfort does not fade. The memory is littered with bits and pieces of images, like a rubbish dump, and it is more and more unlikely that any one form among so many will succeed in standing out.

Global media coverage has plunged us into the era of the "integral catastrophe" that, as Paul Virilio (2005) argues, affects the entire world. Catastrophic images broadcast in real time evokes unprecedented emotions and profoundly influence our perception of reality and the world around us. Media-based stories of traumatic events expand the events' boundaries from geographically constrained to virtually boundless experiences, transforming local events into widespread collective traumas.

The problem lies not in the spread of these images which in themselves can promote knowledge and have transformative power, and furthermore, not releasing them would play into the hands of the deniers. The serious problem lies in the intense, widespread pervasiveness of media propagation. Alarmist information is difficult to working through, and it can increase the likelihood of secondary traumatization causing apocalyptic perceptions and an anticipated solastalgia.

In two empirical studies that investigate the role of visual and iconic representations of climate change in the public domain, Saffron O'Neill and Sophie Nicholson-Code (2009) show that fear-inducing or dramatic representations with the language of alarmism appearing in many guises, e.g. using particularly evocative musical accompaniments to increase tension, are generally an ineffective tool for motivating genuine personal engagement. Tending to disempower and distance people from climate change, they are even a tool of manipulating emotions and creating barriers to participation.

Repeated images lend further impact to the news reports that are in themselves frequently already overloaded and excessive. Although they refer to something familiar to us such as the flow of water or the warmth and glow of fire, they make these events re-emerge as threatening otherness, provoking defensive social reactions of denial fostering cognitive dissonance and distancing dynamics.

Epistemic trust plays a crucial role in human development, shaping individual's ability to regard the knowledge imparted by another person as worthy of credit, generalizable or relevant in its own right, and thus to regard the *other* as a reliable source of knowledge. This adaptive capacity which is fundamental to the phylogenetic development of our species takes shape in the context of early relationships, enabling children to learn from others, and navigate the complexity of the world around them. However, when relationships are problematic or deemed "unsafe," children may develop defensive and self-protective attitudes toward interpersonal trust, leading to two negative outcomes. First, a state of epistemic hypervigilance may occur, characterized by pervasive distrust, leading individuals to reject information from the others altogether. This condition of hypervigilance can stem from experiences of betrayal or inconsistency in early relationships, leading individuals to err on the side of caution by discounting the credibility of others. Conversely, individuals can adopt a stance of credulity in response to unsafe relationships, wherein they indiscriminately accept information from others to avoid conflict that could lead to the breakdown of the bond. This credulity can result from a desire for acceptance or a fear of rejection, leading individuals to overlook critical evaluation of information presented to them. Both epistemic hypervigilance and credulity represent adaptive responses to insecure attachment styles. However, they can hinder individuals' ability to engage critically with information and navigate the complexities of the world effectively. Moreover, these tendencies can extend beyond interpersonal relationships to influence attitudes toward scientific research and knowledge acquisition.

Epistemic distrust, stemming from self-protective simplification in the face of uncertain or dangerous experiences, may manifest as skepticism or rejection of scientific advancements.

Understanding the role of epistemic trust and distrust in shaping attitudes toward knowledge acquisition and interpersonal relationships is essential for fostering healthy development and promoting critical thinking skills. By addressing underlying attachment issues and promoting environments of safety and trust, individuals can cultivate adaptive epistemic attitudes that facilitate open-mindedness and engagement with the complexities of the world.

Catastrophic images obsessively replicated, producing addiction, become an everyday occurrence and no longer cause a scandal. Considering these images as customary and degrading their novelty into obviousness exclude the possibility of working through the consequences of changes. Tali Sharot and Cass Sunstein (2024) say that many people stop noticing what is terrible. They get used to dirty air. They stay in abusive relationships. People grow to accept authoritarianism and take foolish risks. They become unconcerned by their own misconduct, blind to inequality, and are more liable to believe misinformation than ever before.

Mass media behave like those parents who in the mistaken belief that they are protecting their children amplify the dangers and overwhelm them with monstrous and terrifying images frightening them without knowing how to offer them the support and tools to cope with them. Faced with the experience of the new, persecutory states of the mind may be activated, or megalomaniac and omnipotent feelings that equally have the effect of pushing away the possibility of working through the experience (Preta 1999). A suspension of knowledge of the context would be necessary in the face of the impossibility of matching "the imprint and the sight" every time, that is the mental image of the already been and the emerging profile of the here and now (Revelli 2016).

These images encourage blurring of boundaries or an absence of differentiation between actual horror and fiction, fostering growing indifference, desertification of feelings, and technological disenchantment. Over time, these experiences can turn into a sort of emotional illiteracy with an absence of compassion and a loss of a sense of responsibility. The final result can be the collapse of thought in which the distinctions between internal and external reality disappear, replaced by delusional statements of denial and irrational neo-constructions of reality.

Brutal and intrusive images of disasters, when distributed and redistributed millions of times in a continuous stream through traditional and digital media, can become the iconic frame of a decontextualized and sensationalized but also crystallized reality. In this dream-like scenario with a high perceptive invasiveness which captures and hypnotizes, people can evacuate without any elaborative mediation their individual traumatic vicissitudes, their confused feelings of inadequate protection, loss and irreparable damage. Large-scale catastrophes and their anxious experience of ruin, war, natural disasters, and epidemics can become the set in which oneiric experiences of mental precariousness are represented. People could give representation to the abandonment of existential projects, depression and anguish of death, fear of the future, interruption of significant emotional relationships, bereavements and irreparable losses.

If it is true that this dream-like scenario has personal and individual origins, it can also be influenced by societal factors such as serial images. In dreams a person can use images of the environmental disasters as a tool to visualize the darker, less integrated aspects of his personality. An external image, for example one representing a "disaster panorama" at the iconic level, can assume an organizing function in the dream. It facilitates the detailed and organized representation of a set of disorganized internal images in search of representative aggregation. These are the structuring images that appear in dreams or as sensory data in real life and have the quality of containing, of summarizing many of the elements of a particular kind of experience that, when connected, take on a recognizable form, a new visible structure in the image itself. These images, which have a valence of enlightenment, clarification, and condensation of scattered elements can be also defined "epiphanic images" (Correale 2024).

However, in the absence of a stable object, or of a relationship within a container experienced as stable and safe, an external image, a "landscape" proposed at an iconic or generically representative level can act as a tool offering a complex imaginative area, a vividly immersive, almost oneiric theatre for unstable and scattered subjectivity, and risking assuming an organizational function in the face of impulses of an excited and undifferentiated need for identification, which would probably have remained in the shadows.

The object world can be replaced by seductive and threatening internal partial objects that generate confusion and persecution (Williams 2022). Robert Geal (2021) explores films which stage large-scale ecological disasters as enjoyable spectacles of destruction, influencing how the spectator thinks, and therefore how he acts. If a person watches this genre of movies, they repeatedly expose themselves to these threatening images and over time become less emotionally reactive in a sort of desensitization and numbing effect. Viewers might get too used to such images, seeing them as a new normal and being undisturbed by them. These movies are based on recognizing the unfamiliar, the Freudian uncanny, in something familiar. In some cases, binge-watching can be an obsessive and compensatory behavior. This can involve symptoms such as lack of control, negative health and social effects, feelings of guilt, and neglect of duties.

New technologies have made it possible to create and inhabit entire universes of experience that are detached from material and concrete dimensions, to delineate spaces between mind and reality, and to amplify and extend psychic and sensory faculties to excess. These new technologies sometimes functioning as psychic prostheses can make it possible to expand, in an almost unlimited way, the dimensions of an experience that is virtual but at the same time also realistic, thus opening up new fields of experience and mental functioning.

Virtual technology, by blurring the difference between what is objective, what is subjective and what is illusory, has a transforming effect on our thinking. In virtual reality, the image is God, disrupting categories of space and time, connections and conditions on which our subjectivity is built, and as a result, our ideas about reality have to be constantly revised and rearranged to adapt to this new reality (Guerrini Degl'Innocenti 2011).

Virtual reality gives shape, builds the representative frame for confusion sometimes with excited and seductive modes together. It is a frighteningly exciting scenario from which to feel simultaneously attracted and repelled, subjugated in a passivity fed by obscure fascination.

Giuseppe Di Chiara (1999) defines "Psychosocial syndromes" as social situations that can serve as pathological defenses. They constitute forms of collective problematic behavior that is an expression of powerful anxieties of unconscious origin that are shared by society.

I recall in this regard the enactment of so-called doom-scrolling, that is, the hypervigilant tendency to compulsively scroll through the pages of a site, the bulletin board of a social network, searching for information about disasters, catastrophes, loading oneself with emotions that are not one's own and triggering a vicious cycle of self-induced malaise which in the long run one risks not being able to be without. Lucretius seems to be the forerunner of doom-scrolling, when in the proem of the second book of "On the Nature of Things" he describes the sweetness that an observer derives from watching from the shore a ship on the verge of shipwreck.[5]

Obviously, we must avoid naively attributing the origin of experiences of estrangement and the sharp, and pathological separation between "real" and "virtual" solely to media modalities. Analyzing these experiences requires an in-depth understanding of the social and cultural contexts in which these images are produced and consumed.

Ferenczi's Identification with the Aggressor

The concept of "Identification with the aggressor" was developed by Sándor Ferenczi between 1932 and 1933 while working with traumatized patients. It permits us to explore attitudes and beliefs, using psychoanalytical knowledge to help people face the reluctance or passivity to fully acknowledge the seriousness of climate change and change damaging behaviors in our relationship with the environment.

Ferenczi (1933, 1988) first spoke of the identification with the aggressor as the child's response when they feel overwhelmed by threat, losing their sense that the world will protect them, and are in danger with no chance of escape. The child responds with this defense mechanism which plays the part of a protector, producing wish-fulfilling hallucinations and consolation fantasies. This mechanism leads the child to protect himself from unbearable emotions, introjecting the aggressor who disappears as the external reality. Emotional anesthesia of consciousness and consolation fantasies, described by Ferenczi as representing the child's reaction to adult sexual aggression, are a universal tactic of people who find themselves in disproportionately weak positions compared to others seen as powerful and threatening. This can be seen today in our attitudes and responses to climate change (Bellamy 2019).

In order to avoid being emotionally overwhelmed, a mechanism of "auto-hypnosis" may contribute to wholesale climate denial, similar to denial of child abuse. Identification with the aggressor and collusion with the horror followed by an autohypnotic traumatic *trance* permit that the aggressor disappears as external reality and puts us in a delusional, passive "safe place" in which we no longer need anything.

Ferenczi's conception is very different from Anna Freud's later use of the term "identification with the aggressor" in 1936. She denotes how, by impersonating the aggressor, assuming their attributes, or imitating their aggression, the child transforms from the threatened person into the one who makes the threat. Identification with the aggressor constitutes in this way a prodrome of the superego at a normal developmental stage in which aggression is directed outward rather than inward (in the forms of self-accusations, reproaches, criticism, etc.).

Harold Searles (1972) was the first analyst to apply Ferenczi's concepts to the environmental crisis and human vulnerability. He suggests that we unconsciously but powerfully identify with what we perceive as omnipotent technology as a defense against feelings of insignificance and mortality. It is more attractive to indulge in secret fantasies of omnipotent destructiveness, identifying with the forces threatening the environment, than to recognize our own destructive drives, and in this way avoiding to feel guilt and shame.

Ogden's Autistic–Contiguous Position

According to Calvino (1988) today we run the danger in losing a basic human faculty: the power of bringing visions into focus with our eyes shut and in fact of thinking in terms of images. The trauma caused by overly dramatized images can hypersensitize perception, parasitizing and paralyzing psychic functioning, and obstructing the capacity to dream and to play with thoughts, Winnicott's continuous playing between the real object and transitional object (Correale 2021).

Thomas Ogden (1989) describes in the unitary constitution of the self, from the very beginnings of psychological life, the dialectically articulated coexistence of psychological modes of functioning that usually belong to evolutionarily different psychological types and generative modes of experience. He individuates the starting point of mental life (the most primitive psychical structure or state of being) as the "autistic-contiguous" position. He identifies in this way an area of pre-symbolic experience of a sensory nature mainly centered on the surface of the skin. The autistic-contiguous mode is a sensory-dominated, pre-symbolic mode of generating experience which provides a good measure of the boundedness of human experience and the beginnings of a sense of the place where one's experience occurs.[6]

It is to this area that we must refer to in order to describe the relational hypertrophic and rigid sensory modes take over, becoming the prevalent, if not the only form of contact with reality. The characteristic dread revolves around dissolution of sensory boundaries which may be felt as endlessly falling or leaking or otherwise becoming flooded or lost in shapelessness. Rather than using the sensory information to find a way of being (or not being) in relation to the other, the individual gets stuck in a sensory-dominated way of being (or not being). This stops from seeing the symbol and symbolized as separate, leading to a lack of subjectivity.

The attempt to force external reality to homologate to the internal scenario may give rise to anxious-exciting rituals while at the same time constituting a sort of dam against a widespread loss of reality, as would occur in psychotic breakdown. The outside pours into the person through an inflation of visual, where the boundary between animate and inanimate once distinguished by the ancient *pietas* is lost. Pain is not perceived in its essence and the destruction of nature no longer has anything execrable or terrible. The narrative of the macabre develops between patheticism and cynicism; everything is homogenized in a discursive circularity without relevance, where everything becomes equivalent and remains indifferent in the presence of a continuous and endless stream of senseless, or at least interchangeable stimuli. The relentless flood of images not only assumes a pornographic quality but also is obscene, meaning it out of the scene, and erases the human presence from the scene leaving it devoid of humanity. These images can offer a means for disorganized drives in search of representative aggregation to find representation and coherence, providing a sense of structure amidst chaos. They impoverish the cognitive apparatus and the capacity for symbolization by means of symbolic equations similar to those described by Hanna Segal (1957) for psychotic thought, and foster *oneirization* (a dreamlike situation) of the experiential world by feeding essentially emotional and somewhat delusional imaginative constructs.

Another perverse defensive possibility is that the difficulty in connecting with one's own deep-seated anxieties leads to the removal from oneself of any sense of responsibility and any awareness of one's own participation in the creation of the damage, whereby, through exaggerated justifications, one moves from "so much, so everyone does it!" to "so much, everyone likes it!".

In a way, it is as it happens in pedophilic relationships, where the pedophile is convinced that the child is complicit and symmetrically co-responsible for the relationship, that sexual desire is mutual, and the child also enjoys it. The child, matching exactly the excited image the pedophile has of him, wants what he desires. The child with whom the pedophile comes into contact does not have an emotional consistency that is recognized and respected, but is thought of, desired and constructed as a *homunculus*, a kind of disharmonious and artificial miniature adult who, once reified, corresponds exactly to his aroused construction (Schinaia 2010).

In borderline cases, with chaotic and disorganized personalities spread across islands of psychic aggregation, the extensive mass of images proposed by the media, especially those of disasters and catastrophes, distributed with the declared intent to transfix, can act as catalysts, as pseudo-integrative organizers of an internal eroticized confused scenario.

We can witness a regression to pre-symbolic states not only in borderline clinical cases, but also when people, whose identity is still undecided, as happens for instance during adolescence, are bombarded with terrifying images. Individuals are subjected to traumatic stresses because the fast and overwhelming changes of external reality, like what happens to the infant with a manic-depressive mother who continuously changes her mood.

Jackie Finik and Yoko Nomura (2017) had meticulously assembled a research cohort of hundreds of expectant New York mothers. Their investigation – the Stress in Pregnancy study – had aimed since 2009 to explore the potential imprint of prenatal stress on the unborn. Drawing on the evolving field of epigenetics, Finik and Nomura had sought to understand the ways in which the antenatal exposure to environmental stressors could spur changes in gene expression, the likes of which were already known to influence the risk of specific childhood neurobehavioral and psychopathological outcomes.

In late October 2012 superstorm Sandy, a category 3 hurricane, howled into New York City with a force that would etch its name into the annals of history. The storm lent their research a new, urgent question. A subset of their cohort of expectant women had been pregnant during Sandy. Yoko Nomura and her collaborators wanted to know if the prenatal stress of living through a hurricane – of experiencing something so uniquely catastrophic – acted differentially on the children these mothers were carrying, relative to those children who were born before or conceived after the storm.

More than a decade later, the alarming findings of their longitudinal study (Nomura et al. 2022) demonstrate that in utero exposure to a major weather-related disaster was associated with increased risk for psychopathological disorders in children during the preschool years. The researchers suggest paying more attention to understand specific parent, child, and environmental factors which account for this increased risk, and to develop preventive mitigation strategies.

When "inhuman" states of mind cannot be dreamt, or helped to become thought, the annihilation we feel undergoes drastic alteration, pushing us out of contact with our basic animal/human nature which collapses into God-like, circular "thinking" in which distinctions between internal and external realities vanish, replaced by delusional affirmations. Emotional desertification toward the real occurs whenever human beings fail to recognize other human and non-human beings, plants, our natural habitat, as having the status and value of living beings. By suspending the morality that generally governs rational action toward the environment, they cross universal but unfortunately porous boundary, beyond which they become ruthless exterminators without awareness or remorse.

Under such conditions, it becomes possible for normal, morally upright people to commit destructive acts without remorse, without shame, or to witness destructive acts without indignation. We have seen it before and, unfortunately, we are seeing it again during the wars, and also today with regard to Nature.

So, it remains essential to find ways of communicating visually climate change and ecological disasters that have a proactive and transformative value, rather than giving in to the taste for sensationalism and horror. All the stakeholders – including media outlets, policymakers, parents, psycho-analysts, and healthcare professionals – must be aware that the constant exposure to media catastrophic images is likely to have serious and long-lasting consequences.

Conclusion

Freud associates voyeuristic tendencies with a fixation on the primal scene, in which the child witnesses the parents' sexual relationship. This premature traumatic experience can stimulate castration anxiety and drive the child, when he reaches adulthood, to put the scene into practice repeatedly in an attempt to actively overcome a trauma he experienced passively.

Freud (1905, 156) writes: "Visual impressions remain the most frequent pathway along which libidinal excitation is aroused," then looking, supported by the scopophilic drive, is the main way that connects notion of *Schaulust*, "pleasure in looking," in the sense of both seeing and being seen, as well as "curiosity." Freud (1910 [1909]) distinguishes between two frequently encountered forms of this partial drive: one active, "voyeurism," and the other passive, "exhibitionism," neither of which he would necessarily rank among perversions.

According to Stefan Zweig (1942), previous generations could retreat into solitude and seclusion when disaster struck; it was our fate to be aware of everything catastrophic happening anywhere in the world at the hour and the second when it happened. Sontag (2003) sees the contemporary need for media participation in the tragedies of others as a new form of voyeurism.

In these times the opacity of indifference resulting from addiction to serial images, reduces the value of the aesthetic dimension and makes catastrophes capable of spreading across the entire surface of our lives, as one-sided transactions in which we absorb without giving anything back.

There is an "excess of representation" connected to the myriad vivid images that continually invade mental and social space but these, disconnected from the necessary process of symbolization that seems instead lacking, remain like empty shells that multiply in a viral manner without producing nothing but clones of themselves. We have a saturation of the field that prevents critical thinking that would instead need pauses, faults, disarticulations of the real scene (Preta 2023).

More than any other theory and clinical practice, psychoanalysis has the task of comprehending why, in the face of so much evidence of such serious and dangerous damage to us, humankind is unable to understand what has happened and what could still happen. Bion (1974) questions how a human being with a human mentality and personality cannot be concerned with the future. This question needs new analyses where we have a duty to carefully highlight the contextual elements and take into account the complexity we are facing and the need to consider multiple vertices of observation (Lombardozzi 2020).

A critical approach to the crisis obliges us to stay in a dynamic balance as an acrobat who just moves from one imbalance to another in order to balance on the wire. Bion (1973, 1974) advocates a critical capacity – to be able to "think under fire" – a need that holds true today more than ever. In a 1910 letter, Rudyard Kipling wrote to his son:

> "If you can meet with Triumph and Disaster and treat those two imposters just the same [...] yours is the Earth."

In relation to the progressive alteration of the environmental ecosystem, we must avoid the risk both of being pessimistically tempted by catastrophism and looking at novelties as something a priori naively optimistic.

Notes

1 There are four versions of *The Scream* (original title: *Skrik*), the most famous of which (created in 1893 with oil, tempera, pastel and crayon on cardboard, 36 in x 28,9 in) can be found in The National Gallery in Oslo.
2 *El Grito* is a 1978 work by Oswaldo Guayasamín. It is part of the IMMA (Irish Museum of Modern Art) Permanent Collection.
3 The Morandi Bridge has been rebuilt to the design of architect Renzo Piano and was inaugurated in 2020, taking the name Ponte San Giorgio.
4 Solastalgia is a neologism composed by *solace* (comfort, consolation) and *nostalgia* (longing). It is an ecological mourning with the experience of losing a landscape, a

place and a form of identity, which can result in full-blown depression, cynical-negatory defensive modes, impotent passivity and distracted apathy.

5   *Suave, mari magno turbantibus aequora ventis,*
e terra magnum alterius spectare laborem;
non quia vexari quemquamst iucunda voluptas,
sed quibus ipse malis careas quia cernere suave est.
Pleasant it is, when on the great sea, the winds trouble the waters,
to gaze from shore upon another's great tribulation:
Not because any man's troubles are a delectable joy,
but because to perceive what ills you are free from yourself is pleasant.
Lucretius. *On the Nature of Things.* W.H.D. Rouse (Trans), M.F. Smith (Rev). *Loeb Classical Library*, p. 181. Cambridge, MA: Harvard University Press, 1924.

6   Before Thomas Ogden, Esther Bick (1968, 1975) wrote on the containment function of the skin in early object relations and the formation of a "second skin" to describe a replacement mode for a deteriorated sense of cohesiveness of the skin surface. Didier Anzieu (1981) studied the *moi-peau* (skin-ego) formation which has three functions: first, it provides a unitary picture of ourselves; second, it defends us from external intrusions; third, it allows us to relate to other people. Wilfred Bion was more skeptical about the function of the skin and wrote: "My skin is convenient as a method of saying what the boundaries are of my physical make-up, of my anatomy and my physiology. It is unlikely that that forms an adequate description of the boundaries of my mind" (1973, p. 91).

References

Albrecht, G. (2005). "Solastalgia, a new concept in human health and identity". *PAN (Philosophy Activism Nature)*, 3: 41–55.

Albrecht, G. (2020). "Negating solastalgia: An emotional revolution from the Anthropocene to the Symbiocene". *American Imago*, 77(1): 9–30.

Anzieu, D. (1976). "Les esquimaux et les songes". *Revue Française de Psychanalyse*, 40(1): 59–64.

Anzieu, D. (1981). *The Skin-Ego*. N. Segal (Trans). London & New York: Routledge, 2016.

Bellamy, A. (2019). "Trauma, fragmentation and narrative: Sándor Ferenczi's relevance for psychoanalytical perspectives on our response to climate change and environmental destruction". *International Journal of Applied Psychoanalytic Studies: Climate Change and the Human Factor*, 16(2): 100–108.

Bick, E. (1968). "The experience of the skin in early object relations". *International Journal of Psychoanalysis*, 49(2–3): 484–486.

Bick, E. (1975). "Further considerations on the function of the skin in early object relations. Findings from infant observation integrated into child and adult analysis". *British Journal of Psychotherapy*, 2(4): 292–299, 1986.

Bion, W.R. (1973, 1974). *Brazilian Lectures: 1973 São Paulo. 1974 Rio de Janeiro/São Paulo*. London & New York: Routledge, 1990.

Calvino, I. (1988). *Six Memos for the Next Millennium*. P. Creagh (Trans). Cambridge, MA: Harvard University Press, 1988.

Correale, A. (2021). *La Potenza delle Immagini*. Milan: Mimesis.

Correale, A. (2024). "The function of structuring images: The concept of epiphany from literature to psychiatry". *International Review of Psychiatry*, doi:10.1080/09540261.2024.2361760.

Di Chiara, G. (1999). *Sindromi Psicosociali. La Psicoanalisi e le Patologie Sociali.* Milan: Cortina.

Ferenczi, S. (1933). "Confusion of tongues between adults and the child". In *Final Contribution to the Problems and Methods in Psychoanalysis* (pp. 108–125). London: Hogarth, 1955.

Ferenczi, S. (1988). *The Clinical Diary of Sándor Ferenczi.* J. Dupont (Ed.), M. Balint & N.Z. Jackson (Trans). Cambridge, MA: Harvard University Press.

Finik, J. & Nomura, Y. (2017). "Cohort profile: Stress in pregnancy (SIP) study". *International Journal of Epidemiology,* 46(5): 1388–1388k, doi:10.1093/ije/dyw264.

Franckx, C. (2023). "Malaise dans la Nature. La psychanalyse face à la crise climatique". *Revue Belge de Psychanalyse,* 82(1): 111–137.

Freud, A. (1936). *Ego and the Mechanisms of Defense.* C. Baines (Trans). New York: International Universities Press, 1971.

Freud, S. (1905). Three Essays on the Theory of Sexuality. *SE* 7: 123–246.

Freud, S. (1910 [1909]). Five Lectures on Psychoanalysis. *SE* 11: 7–55.

Freud, S. (1919). The Uncanny. *SE* 17: 219–256.

Geal, R. (2021). *Ecological Film Theory and Psychoanalysis: Surviving the Environmental Apocalypse in Cinema.* London & New York: Routledge.

Ghosh, A. (2016). *The Great Derangement: Climate Change and the Unthinkable.* Chicago, IL: University of Chicago Press.

Guerrini Degl'Innocenti, B. (2011). "Spazio onirico e spazio virtuale nel processo analitico". *Rivista di Psicoanalisi,* 57(3): 652.

Havel, V. (1991). *Disturbing the Peace: A Conversation with Karel Huizdala.* New York: Vintage.

Hickman, C. et al. (2021). "Climate anxiety in children and young people and their beliefs about government responses to climate change: A global survey". *Lancet Planet Health,* 5(12): 863–873.

Kassouf, S. (2022). "Thinking catastrophic thoughts: A traumatized sensibility on a hotter planet". *American Journal of Psychoanalysis,* 82(1): 60–79.

Lombardozzi, A. (2020). "Cambiamenti climatici e crisi ambientale. Pensieri psicoanalitici per un'ecologia antropologica". *Rivista di Psicoanalisi,* 66(3): 669–685.

Martellotti, D. (2004). *Architettura dei Sensi.* Rome: Mancosu.

Mirzoeff, N. (2015). *How to See the World.* New Orleans, LA: Pelican.

Nomura, Y., Newcorn, J.H., Ginalis, C., Heiz, C., Zaki, J., Khan, F., Nasrin, M., Die, K., Delngeniis, D., & Hurd, Y.L. (2022). "Prenatal exposure to a natural disaster and early development of psychiatric disorders during the preschool years: Stress in pregnancy study". *The Journal of Child Psychology and Psychiatry,* 64(7): 1080–1091.

O'Neill, S. & Nicholson-Code, S. (2009). "'Fear won't do it': Promoting positive engagement with climate change through visual and iconic representation". *Science Communication,* 30(6): 355–379.

Ogden, T.H. (1989). *The Primitive Edge of Experience.* Northvale, NJ: Jason Aronson.

Pearson, H. (2024). "The rise of eco-anxiety: Scientists wake up to the mental health toll of climate change". *Nature,* 628(8007): 256–258, April 10.

Preta, L. (1999). "L'esperienza del 'perturbante' nell'impatto con le biotecnologie". In L. Preta (Ed), *Nuove Geometrie della Mente* (pp. 5–20). Rome-Bari: Laterza.

Preta, L. (2023). "La fantasia di auto-generazione tra diniego, illusione, speranza". *KnotGarden,* 4: 65–79.

Radua, J., De Prisco, M., Oliva, V., Fico, G., Vieta, E., & Fusar-Poli, P. (2024). "Impact of air pollution and climate change on mental health outcomes: An umbrella review of global evidence". *World Psychiatry*, 23(2): 244–256.

Resnik, S. (1990). *Mental Space*. D. Alcorn (Trans). London & New York: Routledge, 1995.

Revelli, M. (2016). *Non Ti Riconosco. Un Viaggio Eretico nell'Italia che Cambia*. Turin: Einaudi.

Schinaia (2010). *On Paedophilia*. A. Sansone (Trans). London & New York: Routledge.

Schinaia, C. (2019). "Respect for the environment. Psychoanalytic reflections on the ecological crisis". *International Journal of Psychoanalysis*, 100(2): 272–286.

Schinaia, C. (2023). "The uncanny character of images of ecological disasters". In C. Schmidt-Hellerau & M. Erlich-Ginor (Eds), *Mind in the Line of Fire* (pp. 664–667). Columbia, SC: International Psychoanalytical Association.

Schinaia, C. & Todes, K. (2024). "Psychoanalysis and climate change". In G. Gabbard, B.E. Litowitz & P. Williams (Eds), *Textbook of Psychoanalysis* (pp. 49–62). Washington, DC: American Psychiatric Association.

Searles, H.F. (1972). "Unconscious processes in relation to the environmental crisis". *Psychoanalytic Review*, 59(3): 361–374.

Segal, H. (1957). "Notes on symbol formation". *International Journal of Psychoanalysis*, 38(6): 391–397.

Sharot, T. & Sunstein, C.R. (2024). *The Power of Noticing What Was Already There*. New York: Signal.

Simon, P.D., Pakingan, K.A., & Aruta, J.J. (2022). "Measurement of climate change anxiety and its mediating effect between experience of climate change and mitigation actions of Filipino youth". *Educational and Developmental Psychologist*, 39(1): 17–27.

Sontag, S. (2003). *Regarding the Pain of Others*. New York: Farrar, Straus and Giroux.

Van Aken, M. (2020). *Campati per Aria*. Milan: Elèuthera.

Virilio, P. (2005). "Il mondo fragile. Interview with Daniel Biswanger". *Internazionale*, 12(573): 28–32.

Williams, P. (2022). "Rischi nel sopravvivere, rischi nel vivere". *Rivista di Psicoanalisi*, 68(3): 705–723.

Winnicott, D.W. (1974). "Fear of breakdown". *International Review of Psycho-Analysis*, 1 (1–2): 103–107.

Zoja, L. (2011). *Paranoia: The Madness that Makes History*. J. Hunt (Trans). London & New York: Routledge, 2017.

Zweig, S. (1942). *The World of Yesterday: Memoirs of a European*. A. Bell (Trans). London: Pushkin Press, 2011.

Index

on communication through narratives
76–77; on crisis of development
86–87; on function of art 88–89; on
function of the skin 155n6; groups
research 83–84; on importance of
mental health 78; on multiple points
of reference 81; on Nameless Terror
82; on psychoanalysis process 42–43;
on the public factor 43; seeking
asylum in fiction 87
Black Lives Matter campaign 134
Bleger, Josè 6
Bloch, Ernst 10
Boccanegra, Luigi 84
body-as-if-loop theory 84
Bollas, Christopher 79
Bonnefoy, Yves 84
borders, as form of environmental
protection 7–8
Borgna, Eugenio 11
boundaries 147
Bowker, Matthew 2
breakdown, clinical fear of 144
Butler, Judith 36
Byron, George Gordon, "Darkness"
poem 1–2
Byung-Chul Han 45

Cage, John 88
Calvino, Italo 80, 145, 150
Cameron, William B. 6
cannibalistic development 16
carbon economy xv, 59, 62–64
cataclysmic experiences, penetrating core
of mental experience 8
catastrophes: contemporary form of
44–46; etymological meaning of xvi,
75; exposure to today's 44; global
capitalism exploiting 3; of a man's
private existence 47
catastrophic change theory xvi, 104
catastrophism: defined xii; human
development and 109; reductionism as
counter to 18; traumatized
non-dreamer 27; unconscious aspects
of 18
Ceruti, Mauro 14–15, 80, 85
Chakrabarty, Dipesh 94
change, experienced as catastrophic 85
childhood development: challenges of
112–113; direct relationships' impact
on 111; disadvantaged environments

and 111–112; environment's affect on
xvii, 109; inclusive growth 114; as
malleable 111, 112; parenting
practices 113; policy interventions
111–112; see also adolescence
civilization, end of 14–15
climate anxiety 142–143
climate change: deep adaptation to 125;
floods 124–125; food sustainability
and 124–125; literature's relationship
with 46–48; mental health, affects on
142; origins of 93; visual and iconic
representations of 146
climate crisis xiii; accommodating
paradigm shift 64; as agent of
metamorphosis 52; as construction of
subjectivity 53; as a cultural
unthinkable 50–52; as failure of
cultural imagination xv; group defense
mechanisms xviii; infiltrating
adolescent's development process 101;
intergenerational relations and 106;
Oedipal myth and 105; relationships
between generations on 65; species
selection and 4
climate terrorism: anxiety and 5; reacting
to 6–8; risks with xviii–xix
climate warming, provoking sixth
massive extinction 94
collapse 28
collapsology 3–6
collective psychic infections 145
colonialism 93
commedia dell' arte 76, 90n1
complexity, defined 80
composition, overview xii–xiii
concrete hope 10, 82
cooking, natural way of 127–128
Cooper, Steven H. 12
Corrao, Francesco 76
Correale, Antonello 144, 148, 150
counter-Anthropocenes 7
Covid-19 pandemic 68–69, 70–72
creativity 80, 84, 89–90 see also art
credulity 146
crises: consciousness of 56–57;
etymological meaning of 53; as
prognosis of time 54
crisis of adolescence 30
crisis of presence 43–44, 53, 54
culpable man 46
cultural apocalypses 44, 55

For Product Safety Concerns and Information please contact our EU
representative GPSR@taylorandfrancis.com
Taylor & Francis Verlag GmbH, Kaufingerstraße 24, 80331 München, Germany